A bacteriological study of ham souring

Charles Neil McBryde

Alpha Editions

This edition published in 2024

ISBN : 9789366387758

Design and Setting By
Alpha Editions
www.alphaedis.com
Email - info@alphaedis.com

As per information held with us this book is in Public Domain.
This book is a reproduction of an important historical work. Alpha Editions uses the best technology to reproduce historical work in the same manner it was first published to preserve its original nature. Any marks or number seen are left intentionally to preserve its true form.

Contents

INTRODUCTORY. ....- 1 -

METHOD OF CURING HAMS. ....- 3 -

DEFINITION OF SOURING. ....- 6 -

CLASSIFICATION OF SOUR HAMS
AND LOCATION OF SOUR AREAS. ....- 7 -

    METHOD OF DETECTING SOUR
    HAMS. ....- 8 -

THEORIES IN REGARD TO HAM
SOURING. ....- 9 -

PREVIOUS EXPERIMENTAL WORK
TO DETERMINE CAUSE OF HAM
SOURING. ....- 11 -

THE PRESENT EXPERIMENTS. ....- 12 -

    MEDIA EMPLOYED. ....- 12 -

    METHOD OF PROCEDURE IN
    EXAMINING HAMS. ....- 13 -

    RESULTS OF EXAMINATION OF
    SOUR AND SOUND HAMS. ....- 14 -

    HISTOLOGICAL CHANGES IN
    SOUR HAMS. ....- 18 -

CHEMICAL ANALYSES OF SOUR AND SOUND HAMS. - 19 -

BACTERIOLOGICAL EXAMINATION OF SOUR AND SOUND HAMS. - 23 -

INOCULATION EXPERIMENTS WITH HAMS. - 24 -

PROBABLE METHOD BY WHICH HAM-SOURING BACILLUS ENTERS HAMS. - 39 -

BIOLOGICAL AND MORPHOLOGICAL CHARACTERISTICS OF THE HAM-SOURING BACILLUS. - 50 -

PREVENTION OF HAM SOURING. - 60 -

GENERAL SUMMARY AND CONCLUSIONS. - 63 -

ACKNOWLEDGMENTS. - 66 -

INTRODUCTORY.

The souring of hams is a matter of considerable importance to those engaged in the meat-packing industry, and has been the occasion of no little worry, as in even the best-regulated packing establishments the yearly losses it entails are considerable. The subject has given rise to much speculation on the part of those engaged in the curing of meats, as to the cause of the trouble and how it may be remedied, and has received considerable attention in a practical way, but little seems to have been done in a scientific way toward determining the cause and nature of ham souring.

In a well-regulated meat-packing establishment the loss from ham souring is usually figured at about one-tenth of 1 per cent of the total weight of hams cured. At first thought this would seem but a small loss, but when one reflects that in a single large packing establishment some 3,000,000 hams are cured during the year, the loss, when figured out, is considerable. Taking 15 pounds as the average weight of a ham, 3,000,000 hams would represent 45,000,000 pounds of meat. Figuring the loss from souring on the basis mentioned, the amount of meat condemned and destroyed during the year would be 45,000 pounds. Assuming that hams sell at an average wholesale price of 15 cents a pound, the yearly loss for a single plant which cures 3,000,000 hams a year would be nearly $7,000.

While one-tenth of 1 per cent of the total weight of hams cured would represent the loss from souring in a well-regulated establishment, statistics obtained through Government meat inspectors show that 0.25 per cent would more nearly represent the loss for the entire country. During the fiscal year from July 1, 1908, to June 30, 1909, some 670,000,000 pounds of hams were placed in cure in packing establishments subject to Government inspection. Estimating the loss from souring at 0.25 per cent, the total amount of meat condemned and destroyed as sour would be 1,675,000 pounds. At 15 cents a pound the total annual loss from ham souring in packing houses subject to Government inspection would figure up something over a quarter of a million of dollars.

The problem of ham souring, therefore, is quite an important one from a practical and financial standpoint; but aside from these considerations it is also a subject of considerable scientific interest, and in view of the fact that all sour meats are condemned under the Federal regulations governing meat inspection it has seemed fitting that this question should be made the subject of scientific investigation on the part of the Bureau which is charged with the administration of this inspection.

The investigation reported in this paper has been conducted chiefly along bacteriological lines, and has been confined entirely to the wet method of curing hams, as this method is the one generally used in American packing houses.

METHOD OF CURING HAMS.

In order to make clear certain points in regard to the nature and occurrence of ham souring and to insure a better understanding of the experiments which are to be described later, it would seem best to begin with a brief outline of the method of curing hams as practiced in the larger packing establishments of the country. This description is merely a general outline of the method of preparing hams for cure and the method of handling hams while in cure, and deals chiefly with those points that bear on the question of souring.

After the slaughtered animal has been cleaned, scraped, eviscerated, washed, and split down the middle, the carcass is usually allowed to hang for an hour or so in a large room open to the outside air, known as the "hanging floor." This is done with a view to getting rid of a certain amount of the body heat before the carcass is run into the chill rooms, and effects a saving in refrigeration.

The carcasses are next run into "coolers" or chill rooms, and subjected to refrigeration with a view to ridding them entirely of their body heat. The coolers are large rooms fitted with brine pipes and capable of accommodating several hundred carcasses. The temperature of the coolers when the carcasses are run in is about 32° F. When filled, the temperature of the cooler rises to about 45° F., owing to the heat given off from the carcasses. The temperature is then gradually reduced to 28 or 30° F. Hog carcasses are left in the coolers as a rule for forty-eight hours, at the end of which time they are stiff and firm, but not frozen. The temperature of the chill rooms is always carefully watched, thermometer readings being made every few hours and duly recorded. The temperature of the carcasses is always tested when they leave the chill room. In those plants provided with a hanging floor, a certain number of the carcasses are also tested before they are sent to the chill rooms in order to determine the amount of heat lost on the hanging floor.

The carcasses are tested by means of an especially constructed thermometer, known as a "ham thermometer," which has a pointed metal protector so that it can be thrust into the body of the ham. (See fig. 4.) The ham has been rightly selected as the proper portion of the carcass at which to take the temperature, as it constitutes the largest mass of muscular tissue in the carcass and holds the body heat longer than any other portion. In taking the temperature, the thermometer is thrust deep into the body of the ham so that the point of the thermometer rests alongside or a little behind the upper portion of the femur or middle bone, the latter being used as a guide in introducing the thermometer. A certain number of the carcasses from each cooler are tested in this way as a check on the refrigeration. The inside temperature of the hams when they leave the chill rooms should be about 34° F.

The carcasses are next cut up and the hams trimmed for pickling. In some houses the hams are given an additional chilling of 48 hours after they are cut from the carcasses, but this is not done as a rule, nor does it seem to be necessary.

The hams are now sent to the pickling rooms, or "sweet pickle department," as this branch of the packing house is designated, and here a certain number are again tested with a thermometer, as described above. This test is carried out by the foreman in charge of the sweet pickle department in order that he may satisfy himself that the hams are properly chilled before they go into the pickle and as an additional check on the refrigeration.

The hams are now ready to be "pumped," and this pumping, as will be shown later, constitutes an important step in a successful cure. Pumping consists in forcing a strong brine solution containing saltpeter into the muscular tissues of the ham, and is accomplished by means of a large, hollow, fenestrated needle connected by means of a rubber hose with a powerful hand pump. The needle is introduced along the bone, the latter being used as a guide.

In all of the larger packing establishments two general methods of curing hams are followed, the two methods being designated as the "fancy" or "mild cure" and the "regular cure," the term "cure" being used to designate the curing period. Various trade names are given by the different packing establishments to the hams cured by these methods. In the fancy cure the hams are pumped in the shank only, whereas in the regular cure they are pumped in both body and shank. The same pumping pickle is generally used for the two cures. It is a significant fact that the greater proportion of the "sours" are found among the fancy or mild cure hams. This point will be discussed farther on in connection with some experiments to be described later.

The actual curing is usually carried out in large vats which hold about 1,400 pounds of meat or some hundred hams. The hams are packed in the vats in layers and are entirely covered with the pickling solution or brine. A certain proportion is always observed between the weight of the meat and the amount of the solution. The pickling solution, or "pickle," as it is termed, is a brine solution containing saltpeter and sugar. The composition of the pickle varies somewhat with the different packing establishments. The fancy-cure hams are usually cured in a milder pickle, that is, one that contains less salt and saltpeter than the pickle used in the regular cure, although in some packing establishments the same curing pickle is used for the two cures, the only difference being the additional pumping given the regular-cure hams. The pickling rooms, or "cellars," as they are called, are held at a temperature of 34° to 36° F., and the pickling solutions are always chilled to this temperature before being used.

The hams are allowed to remain in cure for about 60 days, and during this time are "overhauled" several times. Overhauling consists in throwing the hams from the vat in which they are packed into a neighboring empty vat, and then transferring the pickle to the new vat. The pickle is not changed, and the same pickle follows the hams through the entire curing process. The object in overhauling is to stir up the pickle and expose fresh surfaces of the meat to its action.

Hams are also cured in tierces which hold about 300 pounds of meat. In the tierce cure, the hams are packed in the tierces, the latter are then headed up, the pickling solution is next run in through the bunghole, so as to fill the tierce entirely, and a wooden stopper is finally driven into the bunghole. The tierces are rolled back and forth across the floor on dates corresponding to the dates of overhauling in the vat cure. The object of the rolling is to stir up the pickle, and in this way it corresponds to overhauling in the vat cure.

DEFINITION OF SOURING.

To the meat inspector, a sour ham is one which has a tainted or "off" odor, that is, any odor which deviates from the normal. The odor may be very slight, so slight that at times only the trained meat inspector can detect it. When slight, the odor is elusive and hard to define, but when pronounced it has a distinctly putrefactive quality. When not very pronounced, the odor possesses, as a rule, a slightly sour quality, chemically speaking, and at times this sour quality may be quite marked; hence the term "sour ham," or "sour" has originated. In a badly soured ham—using the term "sour" in the packing-house sense to denote any ham that is tainted—the odor loses this sour quality and becomes distinctly putrefactive in nature.

CLASSIFICATION OF SOUR HAMS AND LOCATION OF SOUR AREAS.

Sour hams are classed as "shank sours" and "body sours," according to the location of the souring, and these may be "light" or "heavy." When the souring is very pronounced, the ham is termed a "stinker."

Souring appears to start, as a rule, around the stifle joint (femorotibial articulation), and extends upward into the body of the ham.

In quite a large proportion of the hams which are sour in the body—probably from 40 to 50 per cent—the souring extends through to the bone marrow of the femur or middle bone, and the sour odor is at times more pronounced in the bone marrow than in the meat. The odor of the bone marrow, when pronounced, is strongly suggestive of a dissecting-room odor, and is distinctly putrefactive in quality.

In the case of light body sours the sour odor is confined to a small area immediately around the bone, and may be so slight that it is detected only with difficulty. In such hams the bone marrow is apt to be sweet, and it is not until the souring becomes more extensive that the bone marrow becomes involved.

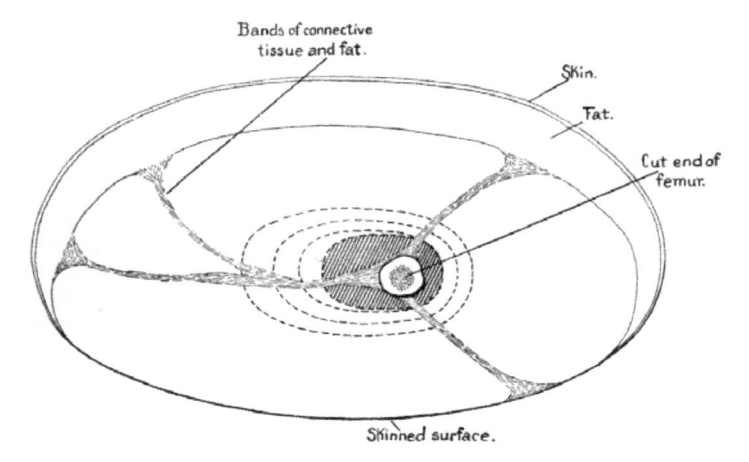

FIG. 1.—Cross section through body of ham, with sour areas indicated by shading and dotted lines.

The distribution of the sour area in the body of a well-developed sour is shown in figure 1.

In the case of a well-developed body sour the sour area is more pronounced near the bone, as represented in figure 1 by the shaded area, and may extend out into the body of the ham for a variable distance, according to the degree of souring, as represented by the dotted lines, gradually fading off toward the margins, where it may be imperceptible or entirely wanting.

In the pronounced sours, termed "stinkers," the odor pervades the entire ham, and is of a distinctly putrefactive quality.

In shank sours, the souring is more or less confined to the shank, or the region about the tibio-femoral articulation, but may extend upward into the lower portion of the body of the ham.

METHOD OF DETECTING SOUR HAMS.

Souring is detected and located by means of a pointed metal instrument known as a "ham trier," which resembles a long, slightly flattened ice pick. The trier is thrust into the ham at different points along the bone, rapidly withdrawn, and the odor which clings to the metal noted. The trained inspector works very rapidly, and is able to detect even the slightest sour or off odor which might be imperceptible to one not trained to the work. At the end of the cure all hams are tested with the trier under the supervision of Government meat inspectors.

Hams are also given what is called the "30-day inspection" by plant inspectors during the process of curing. An average ham weighing from 14 to 16 pounds requires about 60 days to cure, and at the end of 30 days a certain number of hams in each run are usually tested to see how the cure is progressing. If no sour hams are discovered at this inspection the packer knows that the cure is progressing satisfactorily, and moreover he feels sure that his hams will finish satisfactorily, for experience has taught him that souring develops within the first four weeks of the curing period, and if his hams are sweet at the end of this time, he can feel practically sure that no sours will develop later on.

THEORIES IN REGARD TO HAM SOURING.

The theories as to the cause of souring are many and varied. The majority of them are pure speculation and have no foundation upon observed facts. A few of these theories may be enumerated to show how wide and varied has been the speculation upon this subject.

A theory which is quite prevalent among packing-house employees attributes souring to overheating of the animal previous to slaughter, but tests were made by driving hogs to the point of exhaustion just prior to slaughter and curing the hams from these animals in comparison with hams taken from animals which had been rested prior to slaughter, with no difference in the cured product; that is, the hams taken from overheated hogs cured equally as well as those taken from rested hogs.

Another theory attributes souring to a diseased condition of the meat. Prior to the enforcement of the Federal regulations governing meat inspection there might have been some ground for such a supposition, but this theory could not hold at the present time, in view of the thorough and efficient inspection now in force, for it can be safely said that no diseased meat now passes the Government inspectors, and therefore no diseased meat goes into cure in inspected houses. In order to test this theory, however, hams were secured from a number of condemned animals which showed various diseased conditions, such as hog cholera, pyemia, septicemia, scirrhous chord, etc., and these hams were cured in comparison with hams taken from normal hogs. It was found that the hams taken from the diseased hogs cured equally as well as those taken from healthy hogs. The hams from the diseased hogs were destroyed after the experiment, as the meat taken from diseased animals was of course not considered fit for consumption, the object of the experiment being merely to determine whether or not souring is caused by diseased conditions.

Another theory attributes souring to imperfect or too rapid chilling of the meat before it is put in pickle, and places the blame upon the refrigeration. According to this theory, souring results when the meat is chilled too suddenly, the idea being that by the rapid congealing of the juices of the meat a coating is formed on the outside of the ham whereby the animal heat is prevented from escaping from the interior, leaving the meat next to the bone at a higher temperature than the outside of the ham.

In order to test this last theory, a number of hog carcasses were run direct from the killing floor to a cooler at 28° F. and a like number of carcasses of the same average weight which had been allowed to stand for two hours at the outside temperature of the air (53° F.) were placed in the same cooler. The carcasses which had hung for two hours in the air had lost an average of 14 degrees in temperature before going to the cooler. The temperature of the cooler rose to 29° F. after the carcasses were put in, but was soon reduced to 28° F. and held at this temperature. The temperatures of the hams were taken at the end of 24 hours, and practically no difference was found in the inside temperatures of the two lots; that is, the hams on the hot carcasses which were subjected to a sudden chilling exhibited practically the same inside temperature (i. e., next to the bone) as those which had cooled for two hours at the temperature of the air before being placed in the cooler.

Still another theory attributes souring to lack of penetration of the pickling fluids, but analyses of sour and sound hams do not seem to bear out this theory. The rate of penetration of the pickling fluids, however, would seem to have some bearing on the subject, and this point will be discussed later in connection with some laboratory experiments on the inhibitory effects of sodium chlorid and potassium nitrate.

So much for the more commonly accepted theories which have been advanced to explain ham souring.

PREVIOUS EXPERIMENTAL WORK TO DETERMINE CAUSE OF HAM SOURING.

A review of the literature reveals but one article bearing directly on the subject of the cause of ham souring.

In June, 1908, Klein[1] published in the London Lancet an article on "miscured" hams. He describes a miscured ham as one which has a distinctly putrid smell, and the tainted areas he describes as varying in color from a dirty gray to a dirty green, the muscular tissues in the strongly tainted areas being swollen and soft, or jelly-like. From such hams he isolated a large nonmotile, nonspore-bearing, anaerobic bacillus which he calls *Bacillus fœdans*. He cultivated the organism on different media and obtained from the cultures a putrid odor resembling that of the ham from which the culture was obtained, but did not attempt to produce tainting by injecting sound hams with the bacillus.

[1] Klein, E. On the nature and causes of taint in miscured hams. The Lancet, vol. 174, London, June 27, 1908.

While there can be little doubt that Klein's bacillus was the cause of the tainting in those hams which he examined, the proof would certainly have been stronger had he injected sound hams with cultures and thus proven that he could reproduce tainting experimentally by means of his bacillus. Klein examined only dry-cured hams and does not state the temperature at which they were cured. He fails to offer any explanation as to how the bacillus gained entrance into the hams.

THE PRESENT EXPERIMENTS.

MEDIA EMPLOYED.

After considerable experimentation as to a suitable culture medium for the bacteriological study of sour hams, a modification of the "egg-meat mixture" used by Rettger[2] in his studies on putrefaction was found to be the most satisfactory. This medium, which consists of chopped meat and egg albumen, furnishes an excellent medium for the growth of putrefactive organisms which rapidly break down the proteids of the meat, giving rise to the characteristic odors of putrid decomposition. Rettger used chopped beef and egg albumen, but for the present work chopped pork was substituted for the beef, as affording a more suitable medium for the growth of organisms accustomed to growth in pork hams. The modified medium is prepared as follows:

[2] Rettger, L. F. Studies on putrefaction. Journal of Biological Chemistry, vol. 2, 1906.

A. One-half pound of lean pork, freed from excess of fat and sinew, is finely chopped in a meat chopper, 250 cubic centimeters of water is then added, the meat acids are neutralized with sodium carbonate, and the mixture is heated in an Arnold sterilizer for 30 minutes, with occasional stirring. It is then set away in a cold place for several hours. A small amount of fat collects at the top in the form of a fatty scum, as it is impossible to remove all of the fat from the meat before it is chopped. The fatty scum, which hardens upon standing in the cold, is now removed.

B. The whites of three eggs are mixed with 250 cubic centimeters of water. The mixture is rendered neutral to phenolphthalein by means of dilute hydrochloric acid and heated for 30 minutes in the Arnold sterilizer, with occasional stirring.

A and B are now mixed and 2.5 grams (0.5 per cent) of powdered calcium carbonate added. The mixture is next run into large sterile test tubes, or sterile flasks, and sterilized in an Arnold sterilizer on three successive days.

In addition to the egg-pork mixture described above, culture tubes of agar and bouillon prepared from pork instead of beef, with the addition of 1 per cent of glucose, were also used; but the best results were obtained with the egg-pork medium, as with this medium, the early development of sour or putrefactive odors furnished a valuable indication as to the presence of organisms capable of producing sour or putrefactive changes in meat.

METHOD OF PROCEDURE IN EXAMINING HAMS.

The hams were sectioned through the body, the femur, or "middle bone," as it is known in packing-house parlance, being cut at a point about 1-1/2 or 2 inches below its head. A cross section of a ham thus cut is shown in figure 1. After sectioning, the hams were subjected to a microscopical, bacteriological, and chemical examination as follows:

Microscopical examination.—Bits of muscular tissue, taken from various points, were teased out in salt solution and the condition of the muscle fibers noted. Smear preparations were also made from bits of muscular tissue and from the bone marrow, and these were stained and subjected to microscopical examination. Portions of the meat were also hardened and cut into microscopic sections, which were stained and mounted for histological and bacteriological study.

Bacteriological examination.—In the bacteriological examination of sour hams, especial attention was directed to the detection of anaerobic species, as it seemed reasonable to suppose that if the changes taking place in sour hams were due to bacteria these bacteria would in all likelihood be anaerobes (i. e., organisms which develop in the absence of oxygen). This assumption was based upon the fact that, as a rule, souring begins in the interior of the ham next to the bone, and, furthermore, the hams are cured in large vats where they are completely submerged in the pickling fluids, so that any bacteria which develop within the bodies of the hams while they are in cure are probably restricted to practically anaerobic conditions.

Cultures were made from the interiors of the hams at various points by first searing the cut surface thoroughly with a heavy metal spatula and then cutting out, by means of sterile scissors and forceps, plugs of meat about 1 cm. square. The plugs of meat were then dropped into tubes containing the egg-pork medium and pushed down to the bottom of the tubes, where they were held in place by the chopped meat above; in this way conditions favorable for the development of anaerobic organisms were obtained. In inoculating the pork-agar tubes, the medium was first boiled to expel any inclosed air and cooled to 43° to 45° C; the plugs of meat were then dropped into the tubes and the agar rapidly solidified by plunging the tubes in cold water; in this way the bits of meat were inclosed in the agar at the bottom of the tubes, affording suitable conditions for anaerobic growth. Aerobic and anaerobic plates were also made from the meat, and in most cases bouillon tubes were also inoculated. Cultures were always taken from the bone marrow as well as from the meat. Novy jars were also used for obtaining anaerobic conditions in growing the cultures.

Chemical examination.—In order to determine whether the souring was connected with or dependent upon a lack of penetration of the pickling fluids to the interior of the meat, the hams were further subjected to a chemical examination and the content of the meat in sodium chlorid and potassium nitrate determined at varying depths.

RESULTS OF EXAMINATION OF SOUR AND SOUND HAMS.

The sour hams examined were obtained from four different packing establishments. All of the hams studied were "sweet-pickle hams" which had not been smoked. The sour hams selected for examination were good typical body sours, in which the sour odor was well developed, but not of the very pronounced or putrefactive type.

The sour odor in every case was found to be more pronounced next to the bone, being usually rather more pronounced just behind the bone, that is, on the fat side of the bone. The sour odor in each instance was confined to an area of meat immediately surrounding the femur and extending out through the body of the ham for a variable distance, as shown by the dotted lines in figure 1, but in no case did the sour odor extend all the way to the margin of the meat, nor did it as a rule extend below the tibio-femoral articulation, the shank proper and the bone marrow of the shank (i. e., of the tibia) being usually sweet. The butt portion of the hams—that portion above and behind the hitch bone (symphasis pubis)—was also sweet.

Immediately after sectioning, the sour areas, as a rule, could be readily distinguished by a difference in color. In the freshly cut hams the muscular tissue near the bone, where the sour odor was more pronounced, exhibited a slight but distinct grayish hue, at times having a slight greenish tinge; in other words, the muscular tissue in the sour areas lacked the normal bright red color of the sound meat and was distinctly lighter in color than the surrounding tissues. Upon exposure to air, however, the lighter, grayish, sour areas tend to assume a reddish hue and become much less pronounced than in the freshly cut ham. After the cut surface of the ham has been exposed to the air for some time it may be difficult to distinguish the sour areas by any difference in color.

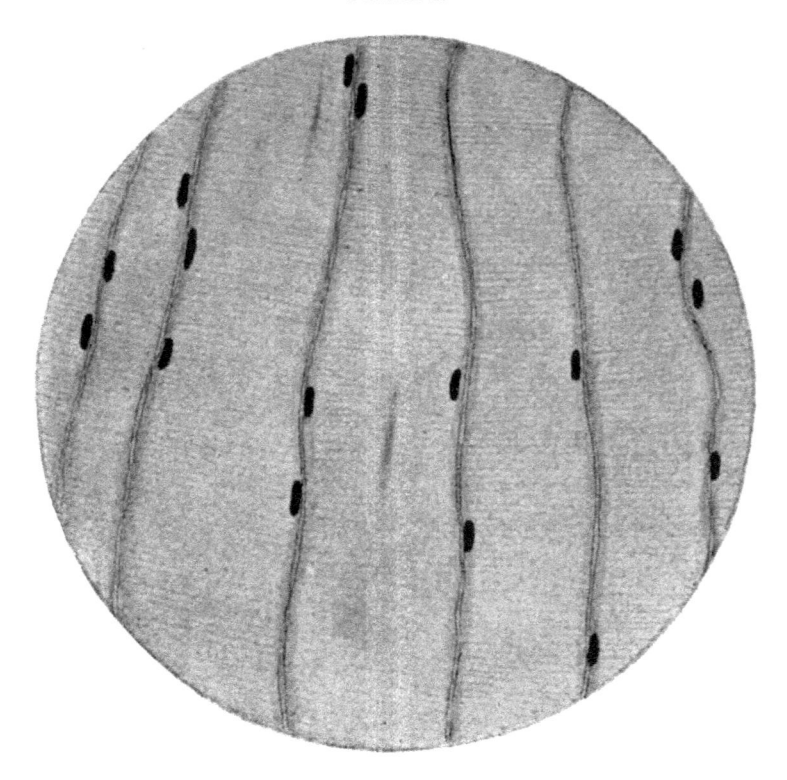

FIG. 1.—SECTION OF MUSCULAR TISSUE FROM SOUND HAM, SHOWING MUSCLE FIBERS CUT LONGITUDINALLY; NUCLEI SHARPLY DEFINED AND CROSS STRIATION DISTINCT.

(Pen-and-ink drawing made with camera lucida from section stained with hematoxylin and eosin to show histological structure.× 320.)

FIG. 2.—SECTION OF MUSCULAR TISSUE FROM SOUR HAM, SHOWING MUSCLE FIBERS CUT LONGITUDINALLY; NUCLEI UNDERGOING DISINTEGRATION AND CROSS STRIATION INDISTINCT.

(Pen-and-ink drawing made with camera lucida from section stained with hematoxylin and eosin to show histological structure.× 320.)

BUL. 132, BUREAU OF ANIMAL INDUSTRY, U. S. DEPT. OF AGRICULTURE.
PLATE II.

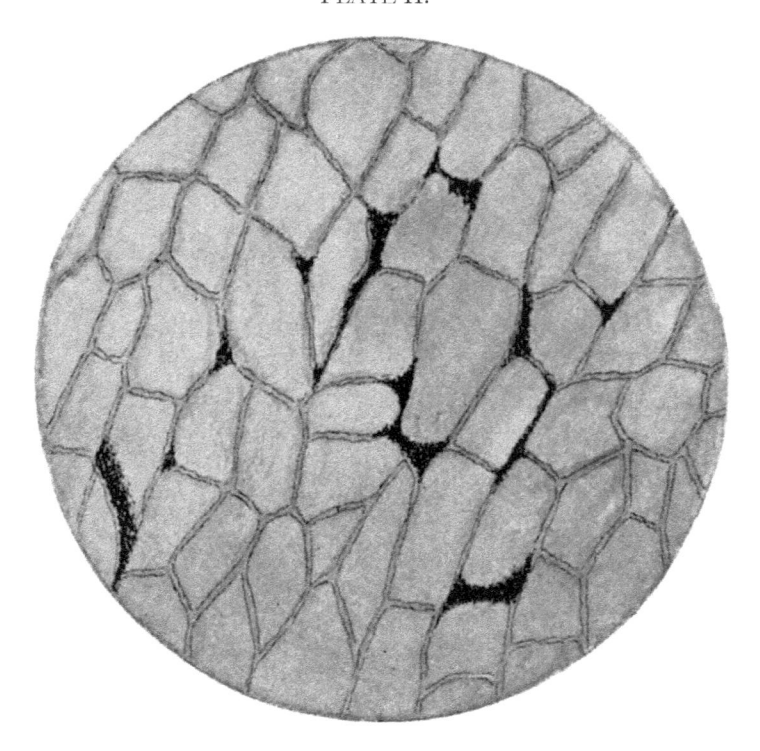

FIG. 1.—SECTION THROUGH MUSCULAR TISSUE OF HAM WHICH HAS UNDERGONE NATURAL OR SPONTANEOUS SOURING, SHOWING DISTRIBUTION OF BACILLI BETWEEN THE MUSCLE FIBERS, WHICH ARE CUT OBLIQUELY. THE DARK MASSES BETWEEN THE MUSCLE FIBERS REPRESENT CLUMPS OF BACILLI.

(Pen-and-ink drawing made with camera lucida from section stained with hematoxylin and eosin to show histological structure.× 320.)

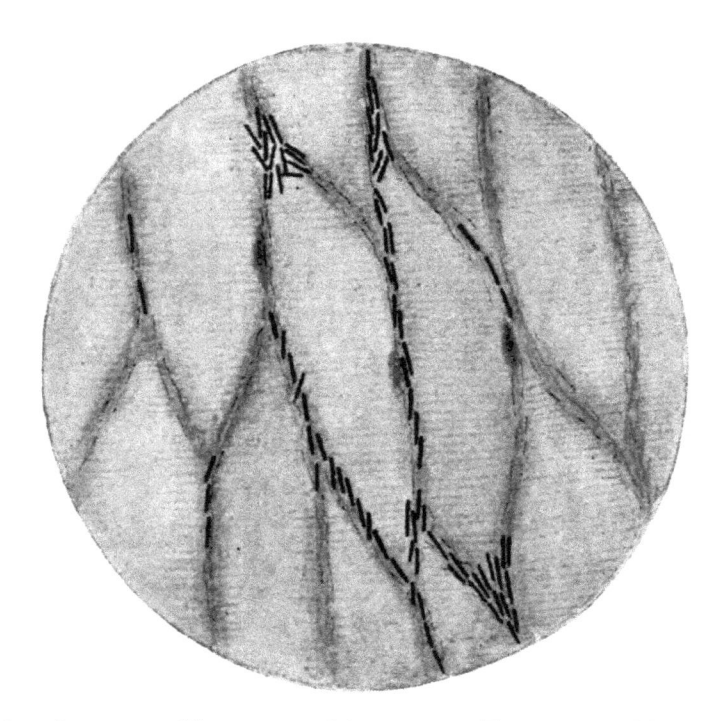

FIG. 2.—SECTION THROUGH MUSCULAR TISSUE OF HAM WHICH HAS UNDERGONE NATURAL OR SPONTANEOUS SOURING, SHOWING INDIVIDUAL BACILLI BETWEEN THE MUSCLE FIBERS, WHICH ARE CUT SOMEWHAT OBLIQUELY. NUCLEI HAVE LOST SHARP OUTLINE AND CROSS STRIATION IS INDISTINCT.

(Pen-and-ink drawing made with camera lucida from section stained with hematoxylin and eosin to show histological structure.× 320.)

In the sour areas near the bone the muscular tissue was distinctly softer; that is, it broke and cut more readily than the surrounding tissues. This was usually quite noticeable in cutting out plugs of the meat for making cultures. In a ham which shows pronounced souring the muscular tissues in the worst affected areas may become quite soft and even slightly gelatinous.

The sour areas, when tested with litmus paper, frequently showed a slight but distinct alkaline reaction. When aqueous extracts of the sour meat, however, were titrated with phenolphthalein they were found to be acid.

HISTOLOGICAL CHANGES IN SOUR HAMS.

In preparations made by teasing out bits of the meat in physiological salt solution, the cross striation of the muscle fibers from the sour areas was found to be much less distinct than in similar preparations taken from sound portions of the meat or from sound hams. At times it was found that the muscle fibers in the sour areas had completely lost their cross striæ, but the longitudinal striation could still be made out. In cases where the souring was pronounced there was sometimes complete loss of both longitudinal and cross striation; in these cases the muscle fibers appeared to have undergone slight swelling and the protoplasm exhibited a finely granular appearance.

In stained sections of the sour meat another striking change was noticed in the disintegration of the nuclei of the muscle fibers, which are at times completely broken up, appearing as bluish granular masses in sections stained with hematoxylin and eosin.(Compare figs. 1 and 2 of Pl. I.)

In sections stained by the Gram-Weigert method to show the presence of bacteria, a large Gram-staining bacillus was noted between the muscle fibers in the connective-tissue elements of the muscle. In some of the sections these bacilli were present in great numbers, sometimes in densely packed clumps or masses, while in other sections, or in other portions of the same section, they were only sparsely distributed between the muscle fibers. Where the bacteria were more numerous the histological changes in the muscle fibers, especially the breaking down of the nuclei, were more noticeable. The intermuscular connective tissue had apparently furnished paths of least resistance along which the organism followed. In Plate II, figures 1 and 2, the bacteria are shown between the muscle fibers under low and high power magnifications.

In Plate II, figure 1, under the low-power magnification, the bacteria appear as dark clumps or bands between the muscle bundles. Under the high power they are shown following along the sarcolemma sheaths between the muscle fibers.

CHEMICAL ANALYSES OF SOUR AND SOUND HAMS.

In order to determine whether there was any difference in regard to the penetration of the pickling fluids in the sour hams as compared with sound hams, a series of four sour hams were subjected to a chemical examination in comparison with four sound hams. All were sweet-pickle hams and were obtained from the same packing establishment. They were all of the same cure and the same approximate age (i. e., length of cure) and the same approximate weight.

In taking samples for chemical analysis, the following procedure was adopted: A section about 2-1/2 inches wide was cut from the center of the body. The two ends of this section were then trimmed off along the lines

L-M and N-O, as shown in figure 2. Beginning at the skinned surface, four slices, A, B, C, and D, were then made, as indicated by the dotted lines. Slice B contained the bone in each instance. Slice D was practically all fat. Each slice was ground separately in a meat chopper and the sample thoroughly mixed before taking out portions for analysis.

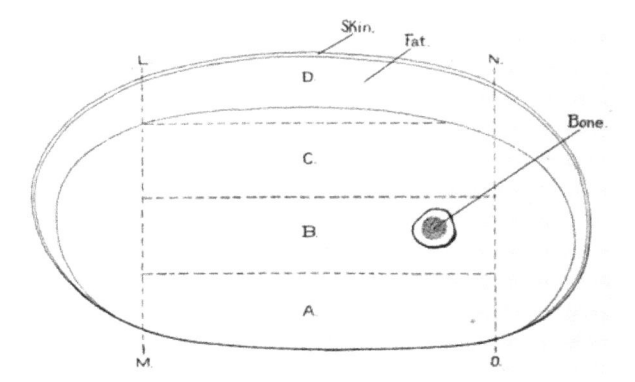

FIG. 2.—Cross section through body of ham to show method of sampling for chemical analysis. A, slice below bone; B, bone slice; C, slice above bone; D, fat slice.

As all of the hams examined were mild-cure hams, that is, had been pumped in the shank only, the pickling fluids in order to reach the bodies of these hams had to penetrate chiefly from the skinned surface of the ham, as little if any penetration takes place through the thick skin of the ham.

The analyses[3] shown in the following tables therefore indicate the degree of penetration of the pickling fluids.

[3] These analyses were made by Mr. R. R. Henley, of the Biochemic Division, Bureau of Animal Industry.

Analyses of sour hams.

No.	Description.	Slice.	NaCl.	KNO$_3$.
			Per cent.	Per cent.
1	Sour body	A	6.18	0.175
		B	4.83	.224
		C	3.65	.299

		D	1.03	.074
2	do	A	5.34	.174
		B	3.70	.150
		C	2.79	.174
		D	1.12	.012
3	do	A	5.04	.125
		B	4.08	.149
		C	2.72	.099
		D	1.19	.048
4	do	A	7.78	.250
		B	5.31	.100
		C	4.76	.200
		D	1.96	.048

Analyses of sound hams.

No.	Description.	Slice.	NaCl.	KNO$_3$.
			Per cent.	*Per cent.*
1	Sound	A	5.80	0.211
		B	4.83	.188
		C	3.86	.221
		D	1.33	.063
2	do	A	4.94	.197
		B	4.08	.149

		C	3.05	.223
		D	1.56	.059
3	do	A	5.92	.173
		B	4.29	.099
		C	4.12	.139
		D	2.32	.049
4	do	A	5.53	.119
		B	4.89	.079
		C	4.32	.099
		D	2.19	.041

Taking an average of the four slices in each ham so as to get an average for the entire ham, and comparing the sour hams with the sound hams, we have the following comparison:

                                                                     NaCl. KNO_3.

Average for 4 sour hams (entire ham)   per cent.   3.84   0.143

Average for 4 sound hams (entire ham)   do   3.93   .131

These figures show practically no difference between the sour and the sound hams as regards the sodium chlorid and potassium nitrate content of the entire ham.

If, now, we compare the bone slices—and these afford really a better basis for comparison, as in sour-body hams the souring is always more pronounced around the bone—we have the following figures:

                                                                     NaCl. KNO_3.

Average for 4 sour hams (bone slice)   per cent.   4.48   0.155

Average for 4 sound hams (bone slice)   do   4.52   0.129

Here, again, we find no essential difference between the sour and the sound hams, and we must conclude from these analyses that souring does not depend upon or result from a lack of penetration of the pickling fluids.

It seems probable that in mild-cure hams, which are pumped in the shank only, the souring begins in the upper portion of the shank and extends upward along the bone into the body of the ham, and that it takes place before the pickling fluid has penetrated to the interior of the ham. When the pickling fluid reaches the interior of the ham it tends to inhibit the souring, which, as will be shown later, is due to the development of bacteria within the bodies of the hams. The growth of the bacteria, however, within the bodies of the hams and the histological changes in the muscle fibers do not seem to interfere with the penetration of the pickling fluids.

BACTERIOLOGICAL EXAMINATION OF SOUR AND SOUND HAMS.

In all of the sour hams which were examined bacteriologically a large anaerobic bacillus was found to be constantly present. From several of the hams this bacillus was obtained in pure culture; that is, it was the only organism present in cultures made from the sour meat and from the bone marrow of the femur. Such cultures, when held at room temperature, gave, at three days, a sour-meat odor exactly resembling that obtained from sour hams.

In many of the sour hams other bacteria were found in association with the anaerobic bacillus noted above. These other bacteria, however, were not constant, being sometimes present and sometimes absent. Among the other bacteria noted in the sour hams, the following forms occurred most frequently:

1. A nonmotile, gram-positive bacillus, measuring from 1.5 to 4 microns in length by 0.5 micron in breadth, sometimes in chains and filaments.

2. A small, nonmotile, gram-negative bacillus, about the size of *Bacillus coli* and usually in pairs.

3. A large micrococcus.

Sometimes one and sometimes all of these bacteria were present in a given ham. They were encountered most frequently in hams which had been pumped in both body and shank, and were probably ordinary pickle bacteria. They were not strict anaerobes, but belonged to the class of facultative or optional anaerobes; that is, organisms which will grow either with or without free oxygen. These bacteria were isolated and grown on the egg-pork medium, but failed to give any characteristic sour or putrefactive odors, and were therefore discarded.

A series of sound hams, all of them of mild cure—that is, hams which had been pumped in the shank only—were also examined bacteriologically. In examining these hams cultures were taken at varying depths, beginning at the skinned surface and going backward toward the fat. Cultures were also taken from the bone marrow of the femur. In the cultures taken near the skinned surfaces the ordinary pickle bacteria were obtained, but these did not, as a rule, extend beyond a depth of 3 centimeters below the skinned surface. The cultures taken from the deeper portions of the hams and from the bone marrow of the femur were entirely negative—that is, failed to show any growth—and the anaerobic bacillus noted in the sour hams was not encountered in any of the cultures made from these hams.

The anaerobic bacillus isolated from the sour hams was found to correspond in morphology with the organism noted in the microscopic sections made from the muscular tissue. In view of this fact and the fact that it was constantly present in the sour hams examined, and was capable of producing in egg-pork cultures a sour-meat odor of the same nature as that obtained from sour hams, this organism was subjected to further study and experimentation.

INOCULATION EXPERIMENTS WITH HAMS.

The experiments which follow were conducted at two different packing establishments in one of the larger packing centers of the country. The officials at each of these establishments showed great interest in the experiments and were most courteous and obliging in supplying the necessary materials.

The first question to be decided was whether the bacillus isolated from sour hams was actually capable of causing ham souring. The bacillus in question had, when cultivated on the egg-pork medium, given rise to a sour odor similar to that obtained from sour hams, but this was not regarded as proof positive that the organism was the actual cause of souring in hams. The proper way to decide this point seemed to be to inoculate hams with the bacillus and then subject these hams to the regular method of cure and see whether they became sour, just as the pathogenic properties of a disease-producing organism are determined by the inoculation of experiment animals. The first two experiments which follow were designed to decide this point.

It was regarded as important to conduct similar experiments at two different establishments, in order to determine whether the same results would be obtained under the somewhat different conditions imposed by different methods of cure. The two experiments which follow were carried out, therefore, at different establishments.

Experiment I.

In carrying out this experiment four tierces of hams were "put down" or "packed"—that is, placed in cure. Two of the tierces were given the fancy or mild cure and two the regular or stronger cure. The hams in two of the tierces, one mild and one regular cure, were injected with a culture suspension of the bacillus; the other two tierces were not injected with culture and were put down to serve as checks on the cure. Hams weighing from 12 to 14 pounds were used for the mild cure, while for the regular cure hams weighing from 14 to 16 pounds were used. This was in accordance with the general rule which prevails in packing houses, the lighter hams being subjected to the mild cure and the heavier hams to the regular cure. The only difference between the mild and the regular cure in this experiment lay in the pumping. The hams which were given the mild cure were pumped in the shank only, while those given the regular cure were pumped in the body as well as in the shank.

All of the hams had received the usual 48-hour chill. They were all pumped with the same pumping pickle and cured in the same curing pickle, and were in cure for the same length of time. The pumping and curing pickles used were the regular pumping and curing pickles of the establishment at which the experiment was carried out, and the hams were cured in accordance with the fancy and regular cures as practiced at this establishment.

The hams were packed in new tierces which had been thoroughly scalded with boiling water. The tierces were held in a curing room which was kept at an average temperature of from 34° to 36° F., the temperature occasionally going as high as 38° and 40° F., but never above 40° F. The hams were left in cure for about 70 days, which is a little longer than the usual cure. The tierces were rolled three times during the cure. At the end of the cure the hams in all four tierces were carefully tested by an expert meat inspector, who knew nothing of the treatment which the hams had received.

The hams in two of the tierces were inoculated with a culture suspension prepared as follows: Ten tubes of egg-pork medium, each tube containing approximately 10 cubic centimeters of the medium, were inoculated with the bacillus and held at room temperature (20° to 25° C.) for six days. The cultures were then filtered through sterile gauze into a large sterile flask; this was done in order to remove the particles of meat, which might otherwise have clogged the syringes used in inoculating the hams. In transferring the contents of the culture tubes to the filter the tubes were washed out with sterile physiological salt solution (0.6 per cent sodium chlorid), and the meat particles on the filter were afterwards washed with the salt solution, a sufficient quantity of the latter being used to bring the total volume of filtrate to 400 cubic centimeters. A microscopic preparation from the filtrate showed the organisms in large numbers, with an occasional rod showing a large terminal spore. This suspension was used for the injection of 40 hams, each ham being given 10 cubic centimeters, or the equivalent of 2.5 cubic centimeters of the original culture. The hams were injected with the culture suspension by means of a sterile syringe carrying a long 5-inch needle. The needle was thrust well into the body of the ham at a point near the upper end of the middle bone or femur, the latter being used as a guide in inserting the needle and the injection being made into the tissues just behind and a little to one side of the upper end of the femur.

The details of the experiment were as follows:

Tierce No. 1 (fancy cure).—This tierce contained 20 hams weighing from 12 to 14 pounds each. These hams were pumped in the shank only. Immediately after pumping they were injected with 10 cubic centimeters each of the liquid culture or suspension described above. After injection the hams were immediately packed in the tierce, which was then headed up, filled with the regular curing pickle, and placed in cure.

Result: When tested at the end of the cure all of the hams in this tierce save one were found to be sour. In 10 of them the souring was very marked throughout the body of the ham and extended into the shank as well. In six the souring was very marked in the body of the ham, but did not extend into the shank. In three there was slight but well-marked souring in the body of the ham with no souring in the shank, and one remained sweet. The probable explanation of the variation in the degree and the extent of the souring will be discussed later. The bone marrow of the femur or middle bone was tested in all of the hams and found to be sour in 18. In one of the hams which showed only slight souring in the body the souring did not extend through to the bone marrow, and in the ham which remained sweet the bone marrow was also sweet. The fact that one ham in this tierce remained sweet was in all likelihood due to an oversight in making the inoculations. In making the inoculations the hams were spread out in a row on a table by a packing-house assistant, who removed the hams as soon as they were inoculated and placed them in tierces; and it is more than probable that the assistant removed one of the hams before it was inoculated in the interval when the writer was busy filling the syringe for the next inoculation.

Tierce No. 2 (fancy cure).—This tierce contained 20 hams of the same average weight as the preceding. They were pumped in the shank only, but were not injected with culture, being put down to serve as checks on the hams in tierce No. 1. These hams, therefore, were subjected to exactly the same cure and were held under exactly the same conditions as those in tierce No. 1, the only difference being that the hams in this tierce were not injected with culture.

Result: When tested at the end of the cure all of the hams in this tierce were found to be perfectly sound and sweet, showing that the curing in this instance was properly carried out and that the souring of the hams in tierce No. 1 was undoubtedly due to the injections of culture which they received.

Tierce No. 3 (regular cure).—This tierce contained 20 hams weighing from 14 to 16 pounds each. These hams were pumped in the shank and also in the body. Immediately after pumping they were each injected in the same manner as those in tierce No. 1 with 10 cubic centimeters of culture. The hams were then packed in tierce and placed in cure.

Result: At the end of the cure 9 of the hams were found to be sour, while 11 remained sweet. Of the 9 hams which became sour, 1 showed very pronounced souring in the body and in the shank as well, 3 showed very pronounced souring in the body, 1 showed pronounced souring in the body, and 4 slight souring in the body. The bone marrow of the femur was tested in all of the sour hams and was found to be sour in 7. In 2 of the sour hams which showed slight souring in the body the souring noted in the meat had not extended through to the bone marrow.

Tierce No. 4 (regular cure).—This tierce contained 20 hams of the same average weight as those in tierce No. 3, and, like the latter, were pumped in both shank and body, but were not injected with culture. This tierce was put down to serve as a check on tierce No. 3 and was held under exactly the same conditions, the only difference being that these hams were not injected with culture.

Result: At the end of the cure the hams were carefully tested and all were found to be perfectly sound and sweet.

Results of Experiment I.

No. of tierce.	Number of hams.	Average weight of hams Pounds.	Cure.	How pumped.	Treatment.	Number of sour hams	Percentage of sour hams
1	20	12-14	Fancy	Shank	Each ham	19	95

				only	injected with 10 c. c. of culture.		
2	20	12-14	do	do	Not injected with culture; check on tierce 1.	0	0
3	20	14-16	Regular	Shank and body	Each ham injected with 10 c. c. of culture.	9	45
4	20	14-16	do	do	Not injected with culture; check on tierce 3.	0	0

Three hams from each tierce were selected for bacteriological and histological examination. From tierces 1 and 3, which contained the injected hams, three of the most pronounced "sours" were selected from each tierce. In examining the hams bacteriologically the following method was adopted: The hams were sectioned near the center of the body and the larger or butt end turned up so as to expose the cut surface. A cross section of a ham thus cut is shown in figure 3.

Cultures were taken at the points indicated by the numbers and from the exposed bone marrow of the femur by first searing the surface, and then taking out plugs of the meat or marrow by means of sterile instruments. The plugs of meat or marrow were dropped into tubes containing egg-pork medium and pushed to the bottom of the tubes by means of a sterile platinum wire. In the cultures made from the sour hams from tierces 1 and 3, which were injected with culture, the bacillus with which these hams were injected was found in practically every culture, although it was sometimes absent in the cultures taken at points near the skinned surfaces of the hams (i. e., at points 1, 4, and 5 in fig. 3). In the cultures taken from the meat, the bacillus was not always present in pure culture, but this is not to be wondered at when we remember that the pickling fluids often contain large numbers of bacteria of various kinds, and these, of course, find their way into the hams in the pickling fluids. Especially is this true of the hams which are pumped in the body, where bacteria are actually pumped into the bodies of the hams in the pumping pickle. In the case of hams which are not pumped in the body, the pickle bacteria do not appear to penetrate the body of the ham to any great depth.

In figure 3 the plus signs after the figures represent the distribution of the sour-ham bacillus in one of the hams from tierce 1, and this may be taken as a typical example of the other sour hams which were examined in this experiment. It should be explained that the shaded areas are not intended to represent the actual limits of souring, but simply the areas in which the sour odor was most pronounced and from which it could be readily obtained with the trier. In comparing the regular and mild cure hams, it was found that the areas of souring as defined with the trier were more restricted in the regular cure hams, and this was undoubtedly due to the additional pumping which these hams received, whereby the growth of the bacillus was partially inhibited.

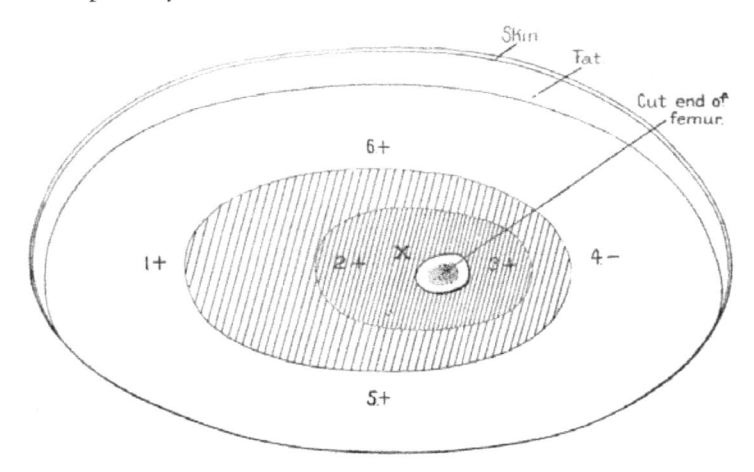

FIG. 3.—Cross section through body of artificially soured ham, showing sour areas and points at which cultures were taken. Darker shading indicates sour area in hams pumped in body and shank; light shading indicates sour area in hams pumped in shank only; figures indicate points at which cultures were taken; plus signs indicate presence of bacillus; minus sign indicates absence of bacillus; X indicates point of inoculation.

It will be noticed that the sour-ham bacillus was present in cultures taken at points outside the shaded areas, indicating that the organism had extended generally throughout the bodies of the hams. As the hams were inoculated at a point just to one side of and a little behind the femur (i. e., at the point X in the figure), the presence of the bacillus generally throughout the hams would indicate a very extensive multiplication of the original bacilli with which the hams were injected. In view of the fact that the bacillus in question is nonmotile, the spread of the bacilli throughout the hams must result simply from subdivision and growth by extension, and in spreading throughout the hams the bacilli appear to follow along the connective tissue bands which afford paths of least resistance. In the cultures made from the bone marrow the bacillus was recovered in pure culture from each of the hams examined, and it is probable that the bacillus finds its way into the bone marrow from the meat by following along the small arteries which pass through the bone. The fact that the bacillus was found in pure culture (i. e., uncontaminated) in the cultures made from the bone marrow is explained probably by its capacity for growth by extension, and also by the fact that the pickling solutions probably do not reach the bone marrow until late in the curing and then only to a limited extent. The bacteria which ordinarily occur in pickling fluids are not strict anaerobes and are not placed under the most suitable conditions for growth when they reach the interior of the ham, for it seems probable that in the interior of hams which are totally submerged in pickling fluids the amount of available oxygen must be extremely small. The ordinary pickle bacteria, therefore, would not multiply as rapidly in the interior of the hams and would not find their way into the bone marrow as soon as would a strictly anaerobic organism.

Pure cultures of the sour-ham bacillus, recovered from the meat and bone marrow of the injected hams, were compared with cultures of the original bacillus used for inoculating the hams, and were found to be identical. Furthermore, the bacillus with which the hams were injected was recovered from the injected hams at points far removed from the original point of injection, showing that the organism had multiplied and extended throughout the bodies of the hams and that it was clearly responsible for the souring which the hams had undergone.

Sound hams from tierces 2 and 4 were examined bacteriologically in the same manner as the injected hams, and some of the cultures showed the ordinary pickle bacteria, but in not a single instance did egg-pork cultures yield a sour odor, and in no case could the sour-ham bacillus be demonstrated in any of these hams.

Microscopic sections and teased preparations of the muscle fibers in salt solution were prepared from several of the sour hams in this experiment, and these preparations showed the same histological changes and the same distribution of bacilli as noted in the natural sours.

In [Plate III](), figures 1 and 2, sections are shown of artificially soured hams, that is, hams which were artificially soured by injections of culture; and if these figures be compared with the sections made from hams which had undergone spontaneous souring (see Pl. II, figs. 1 and 2) the similarity in the form and distribution of the bacilli will be at once apparent.

BUL. 132, BUREAU OF ANIMAL INDUSTRY, U. S. DEPT. OF AGRICULTURE.
PLATE III.

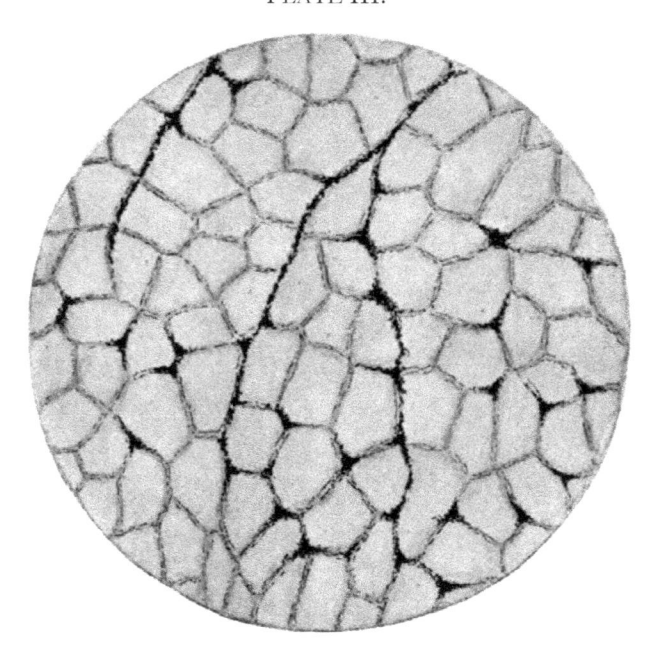

FIG. 1.—SECTION THROUGH MUSCULAR TISSUE OF ARTIFICIALLY SOURED HAM, SHOWING DISTRIBUTION OF BACILLI BETWEEN THE MUSCLE FIBERS, WHICH ARE SHOWN IN CROSS SECTION. THE DARK LINES AND MASSES BETWEEN THE MUSCLE FIBERS REPRESENT CLUMPS OF BACILLI.

(Pen-and-ink drawing made with camera lucida from section stained by the Gram-Weigert method to show bacteria. × 85.)

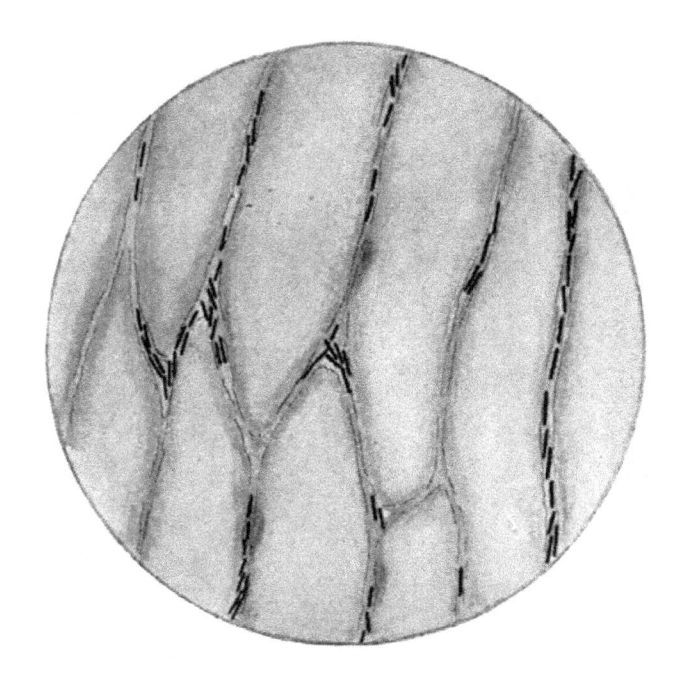

FIG. 2.—SECTION THROUGH MUSCULAR TISSUE OF ARTIFICIALLY SOURED HAM, SHOWING INDIVIDUAL BACILLI BETWEEN THE MUSCLE FIBERS, WHICH ARE CUT LONGITUDINALLY.

(Pen-and-ink drawing made with camera lucida from section stained by the Gram-Weigert method to show bacteria. × 320.)]

Summary and discussion of Experiment I.—Comparing tierces 1 and 2, where the hams were pumped in the shank only, the only difference being that the hams in tierce 1 were inoculated with culture while those in tierce 2 were not, we find that in tierce 1 nineteen out of twenty, or 95 per cent, of the hams became sour, whereas in tierce 2 all of the hams remained sweet. In view of the fact that these tierces were held under exactly the same conditions, we must conclude that the souring of the hams in tierce 1 was due to the injection of culture which they received.

Comparing tierces 3 and 4, where the hams were pumped in both shank and body, the hams in tierce 3 being injected with culture while those in tierce 4 were not, we find that in tierce 3 nine out of twenty, or 45 per cent, of the hams became sour, whereas in tierce 4 all of the hams remained sweet. As the conditions of cure were the same for all four tierces, we must again conclude that the souring of the hams in tierce 3 was directly attributable to the injections of culture which they received.

If now we compare tierces 1 and 3, the two tierces which were injected with culture, we find that in the case of tierce 1, where the hams were pumped in the shank only, 95 per cent became sour; whereas in the case of tierce 3, where the hams were pumped in both shank and body, only 45 per cent became sour. In other words, the percentage of souring in those hams which were pumped in the body as well as in the shank was 50 percent less than in those hams which were pumped in the shank only. Inasmuch as the only difference in the treatment accorded tierces 1 and 3 lay in the additional pumping given the hams in tierce 3, we must conclude that the marked diminution in the percentage of souring in the case of tierce 3 was undoubtedly due to the additional pumping which these hams received, the hams being saturated at the start with the pumping pickle. It will be shown later that both sodium chlorid and potassium nitrate exert an inhibitory effect upon the bacillus with which the hams were injected, which directly bears out the foregoing conclusion.

In tierces 2 and 4, the two check tierces which were not injected with culture, all of the hams were sweet at the end of the cure, showing that the conditions under which the experiment was carried out were entirely favorable to a successful cure.

The sour odor obtained from the artificially soured hams in this experiment was pronounced by the meat inspector who tested the hams, and who was entirely unaware of the treatment they had received, to be identical with the usual sour odor which characterizes hams that have undergone spontaneous souring; in other words, there was no difference in odor between these artificially soured hams and natural sours.

With regard to the variation in the degree and the extent of the souring exhibited by the individual hams in the two inoculated tierces, where some of the hams showed pronounced souring throughout the body and shank, while others which had been injected with the same amount of culture showed only slight souring in the body, several factors must be considered, viz:(1) Differences in the reaction of the meat of the individual hams which may have exerted an influence on the growth of the bacteria with which the hams were injected. (2) Variations in the texture of the muscle fibers and connective tissue of the individual hams, permitting in some cases a more rapid and thorough penetration of the pickling fluids to the interior of the hams, whereby the inhibitory effect of the sodium chlorid and the potassium nitrate on the bacteria would come into play earlier. (3) Variations in pumping, whereby more of the pickling solution was forced into some of the hams than into others. Probably all three of these factors would have to be taken into account in explaining the variation in the degree and extent of the souring exhibited by the injected hams.

With regard to the souring of the bone marrow, we find that of nineteen sour hams in tierce 1 eighteen showed sour marrows, while in tierce 3 nine sour hams showed seven sour marrows. The high proportion of marrow sours is not surprising when it is recalled that of the nineteen sour hams in tierce 1 the meat was markedly sour in sixteen, while of the nine sour hams in tierce 3 the meat was markedly sour in five. In the case of the four sour hams in tierce 3 which showed slight souring in the body, two of these showed sour marrows, while in two the marrows were sweet. In this experiment the percentage of sour hams showing sour marrows corresponds with the percentage of marrow-sour hams found in the packing house, where, as has been pointed out before, a ham which is markedly sour in the body will practically always show sour marrow, while in hams which show only slight souring in the body the marrow is involved in about 50 per cent of the cases.

Experiment II.

This experiment was essentially a repetition of Experiment I, but was carried out at a different packing establishment and under somewhat different conditions.

Two lots of hams were injected with a culture suspension of the bacillus at different stages of the cure, or rather at different stages in the preparation for cure, i. e.,(1) on the hanging floor, previous to chilling, and (2) after chilling and pumping and immediately before packing. Three tierces, each containing 20 hams, were put down. Two of the tierces contained the hams injected with culture, while the third tierce contained check hams which had not been treated with culture. Half of the hams in each tierce were pumped in the shank, while the other half were pumped in both body and shank. The same pumping and curing pickles were used for all three tierces, and were the regular pumping and regular curing pickles of the establishment at which the experiment was carried out. The hams used were all 14 to 16 pounds in weight and were subjected to the usual 48-hour chill with an additional chill of 48 hours after they were cut from the carcass. They were packed in tierces which had been thoroughly scrubbed and cleaned with boiling water. The tierces were held in a pickling room at a temperature of 33° to 36° F., the temperature never rising above 36° F., and were rolled three times during the curing period. The hams were in cure for about eighty days. At the end of the cure the hams were carefully tested by a trained meat inspector, who knew nothing of the treatment they had received.

The culture suspension was prepared from 20 tubes of egg-pork medium in the same manner as that used in Experiment I, the cultures being diluted with sufficient salt solution to give 400 cubic centimeters of suspension.

The cultures from which the suspension was prepared had grown at room temperature for ten days. The suspension was examined microscopically and showed large numbers of the bacilli in the form of filaments or long chains, with many of the individual organisms showing large terminal spores. The hams were injected with the culture suspension in the same manner as those in Experiment I.

The details of the experiment were as follows:

Tierce No. 1.—Contained 20 hams, each ham being injected with 20 cubic centimeters of the suspension or the equivalent of 10 cubic centimeters of the original culture. The hams were injected while on the hanging floor, before they had been cut from the carcasses and previous to chilling. The carcasses were still quite warm—that is, had lost but little of their body heat when the injections were made. The carcasses, which had been carefully tagged, were then run into coolers and given the usual 48-hour chill, after which the hams were severed from the carcasses and given an additional 48-hour chill in accordance with the custom of the packing house at which the experiment was carried out. The hams were next pumped with regular pumping pickle, 10 being pumped in both body and shank and 10 in shank only. They were finally packed in a tierce, which was then headed up, filled with regular curing pickle, and placed in cure.

Result: When tested at the end of the cure it was found that the 10 hams which were pumped in the shank only were all sour. In each of them the souring extended throughout the entire ham, in the shank as well as in the body, and was very pronounced, so much so that they were characterized as "stinkers" by the meat inspector who assisted in testing them. The bone marrow of the femur or middle bone was sour in all of these hams. Of the 10 hams which were pumped in both body and shank 7 showed well-marked souring throughout the body, but the souring did not extend into the shank. The bone marrow of the femur was found to be sour in 6 of these hams, while in 1 the souring had not extended through to the bone marrow.

Tierce No. 2.—Contained 20 hams which were chilled and pumped in exactly the same manner as those in tierce No. 1. These hams were injected with culture after they had been chilled and pumped, or just before they were placed in cure. The hams in this tierce, therefore, were injected with culture four days later than those in tierce 1. The hams were injected with a bacterial suspension prepared in the same manner as that used for tierce 1, except that the egg-pork cultures from which the suspension was prepared were 7 days instead of 10 days old. Each ham was injected with 20 cubic centimeters of the suspension or the equivalent of 10 cubic centimeters of the original culture. The hams were injected in the same manner as those in tierce 1.

Result: When tested at the end of the cure, it was found that of the 10 hams which were pumped in the shank all were sour; in 8 of these the souring was very marked throughout the body of the ham and extended into the shank; in all of these hams the souring had extended through to the bone marrow of the middle bone or femur. Of the 10 hams which were pumped in both body and shank 6 were sour in the body. These hams were classed by the meat inspector who examined them as "light body sours," and in none of them did the souring extend into the shank or through the bone into the bone marrow of the femur.

Tierce No. 3.—Contained 20 hams which were chilled and pumped in the same manner as those in the two preceding tierces. These hams were not injected with culture and were put down to serve as checks on the cure. In other words, they were pumped with the same pickling fluids, were subjected to exactly the same cure, and were held under precisely the same conditions as those in the preceding tierces, the only difference being that the hams in this tierce were not injected with culture.

Result: When tested at the end of the cure, all of the hams in this tierce were found to be perfectly sound and sweet.

Results of Experiment II.

No. of tierce.	Number of hams.	Average weight of hams. Pounds.	How pumped.	Treatment.	Number of sour hams.	Percentage of sour hams.
1	20	14-16	10 in shank	Each ham injected with 20 c. c. of culture prior to chilling and pumping.	10	100
			10 in body and shank	do	7	70
2	20	14-16	10 in shank	Each ham injected with 20 c. c. of culture subsequent to chilling and pumping.	10	100

			10 in body and shank	do	6	60
3	20	14-16	10 in shank	Not injected with culture	0	0
			10 in body and shank	do	0	0

Four hams were selected from each tierce for bacteriological and histological examination. From tierces 1 and 2, in which the hams were injected with culture, 4 of the sourest hams were selected from each tierce. Cultures were made from these hams in the same manner as described under Experiment I and with the same result—that is, the sour-ham bacillus was found throughout the bodies of the hams. Microscopic sections were also prepared from these hams and showed the same histological changes and the same distribution of bacilli as noted for the hams in Experiment I.

Summary and discussion of Experiment II.—Comparing tierces 1 and 2, in which the hams were injected with culture, with tierce 3, where the hams were not injected with culture, we find that in tierce 1 seventeen hams (85 per cent) became sour and in tierce 2 sixteen hams (80 per cent) became sour, whereas in tierce 3 all of the hams were sweet. The fact that all of the hams in tierce 3, the check tierce, were sweet indicates that the conditions were favorable for a successful cure; and as all three tierces were cured under exactly the same conditions, the only difference being that the hams in tierces 1 and 2 were injected with culture, whereas those in tierce 3 were not injected with culture, we must conclude that the souring of the hams in tierces 1 and 2 was due to the injections of culture which they received.

Comparing tierce 1 with tierce 2, we find that the hams in tierce 1 showed more extensive souring than did those in tierce 2, this being especially noticeable in the case of the hams which were pumped in both body and shank. This difference in the extent or degree of souring was probably due to the fact that the hams in tierce 1 were injected while they were still warm and before they had lost their animal heat, the bacterial suspension thus having a better chance to become disseminated through the meat. The hams in tierce 2 were injected with culture after they had been chilled, when the tissues were more or less contracted and the conditions less favorable for the dissemination of the suspension throughout the meat. The hams in tierce 1 were also injected four days earlier than those in tierce 2, and prior to pumping; and this would explain the greater difference in the extent of the souring in the case of the hams which were pumped in both body and shank, as in tierce 1 the bacteria had four days in which to develop before coming in contact with the pickling fluids, whereas in tierce 2 the bacteria were injected after the hams were pumped with pickle and were thus brought into immediate contact with the pickling fluids, which, as will be shown later, have a distinct inhibitory action upon the bacillus in question. In the case of the hams which were pumped in the shank but not in the body there was not this difference, as in these hams the pickling fluids must penetrate into the bodies of the hams from the outside. As it requires some time for the pickling fluids to reach the interior of a ham, the bacteria were thus afforded quite an interval in which to develop before being exposed to the inhibitory action of the pickling fluids. A chemical study of the processes involved in ham curing has been carried out in the Biochemic Division and the approximate rate of penetration of the curing pickle determined, and it was found that it required about four weeks for the interior of a 10-pound ham which had not been pumped to acquire its maximum percentage of sodium chlorid.

To recapitulate: In this experiment 40 hams were injected with culture, half of this number being pumped in the shank only and half in both body and shank. Of the 20 which were pumped in the shank only, every ham without exception, or 100 per cent, became sour. Of those which were pumped in both body and shank, 13, or 65 per cent, became sour. The reduction in the percentage of sours in the last lot was clearly due to the additional pumping which these hams received.

If now we compare tierce 2 in this experiment with tierces 1 and 3 in Experiment I—these three tierces being comparable, as they were all injected with culture at the same stage in their preparation for cure, that is, subsequent to chilling and pumping—we find, in the case of the hams pumped in both body and shank, 65 per cent of sours in Experiment II as against 45 per cent in Experiment I, and this difference is undoubtedly due to the heavier dose of culture used in Experiment II, where the hams were given the equivalent of 10 cubic centimeters of egg-pork culture as against 2-1/2 cubic centimeters in Experiment I. In the case of the hams which were pumped in the shank but not in the body, the percentage of sours was practically the same in the two experiments—in Experiment I all but one of these hams became sour, while in Experiment II all of them became sour. The degree or extent of the souring in these last hams, however, was greater in Experiment II, a result of the heavier injections of culture which they received.

Summary of Experiments I and II.

Summarizing the results obtained in Experiments I and II, we find that culture suspensions of the anaerobic bacillus isolated from sour hams caused souring with great uniformity when injected into the bodies of sound hams which were pumped in the shank only. In the two experiments, 40 sound hams which were pumped in the shank only were injected with culture suspensions of the bacillus, with the result that 39, or 97.5 per cent, became sour during the process of cure; and it is quite probable, as we have pointed out before, that one of these hams was overlooked in making the inoculations, otherwise the entire lot would have become sour.

The inhibitory action of the pickling fluids upon the bacillus is well shown in the case of those hams which were pumped in both body and shank. Out of 40 hams which were pumped in both body and shank, 22, or 55 per cent, became sour in the process of curing. Inasmuch as these hams were cured under precisely the same conditions as the hams which were pumped in the shank only, we must conclude that the diminution in souring in these hams was undoubtedly due to the additional pumping which they received, whereby the bacteria with which these hams were injected were brought into immediate contact with the strong pumping pickle and their development thereby inhibited.

In these two experiments it was proven beyond doubt that the anaerobic bacillus isolated from sour hams was capable of producing souring when introduced into the bodies of sound hams; and in view of the fact that this bacillus was constantly present in hams which had undergone spontaneous or natural souring, and was the only organism that could be isolated from such hams that was capable of producing in egg-pork cultures the characteristic sour-ham odor, the conclusion seems justifiable that this bacillus is an undoubted cause of the ham souring which occurs in the packing house; and the results thus far obtained indicate that it is an important, if not the only, factor concerned in ham souring.

Having established the etiological relation of the bacillus isolated from sour hams with ham souring, the next point to be considered was the manner in which this bacillus finds its way into the bodies of the hams.

PROBABLE METHOD BY WHICH HAM-SOURING BACILLUS ENTERS HAMS.

Regarding the question of the probable method by which the ham-souring bacillus enters hams, there were three possibilities to be taken into consideration:(1) That the bacillus is present in the flesh of hogs at the time of slaughter,(2) that the bacillus gains entrance through the pickling fluids,(3) that the bacillus is introduced into the bodies of the hams in the handling or manipulation which the hams undergo in preparation for, or during, the process of curing.

POSSIBILITY OF INFECTION PRIOR TO SLAUGHTER.

In order to throw some light upon this point, a number of fresh hams—that is, hams which had been chilled but not pumped or subjected to any other manipulation—were examined bacteriologically, but in no case could the anaerobic bacillus which was isolated from sour hams be detected in any of them. The fact that in certain of the smaller packing establishments which cure their hams without pumping the percentage of souring is extremely low would also seem to negative this possibility, for if the bacillus which causes souring were present in the hams at the time of slaughter, sour hams would be as frequent at such establishments as at those establishments which make a practice of pumping. Furthermore, a laboratory study, biological and chemical, of the bacillus isolated from sour hams shows that this organism belongs to the class of putrefactive bacteria, and while such bacteria may be present in the intestines of healthy animals, as, for example, the bacillus of Bienstock (*Bacillus putrificus*), these bacteria do not invade the organs and tissues of the body until after the death of the animal, and the packing-house practice of rapidly eviscerating the hogs immediately after slaughter would certainly preclude this possibility.

POSSIBLE INFECTION FROM PICKLING FLUIDS.

With regard to the second possibility, that the bacillus finds its way into the hams in the curing pickles, it was determined by laboratory experiment that the addition of 3 per cent of sodium chlorid or 3 per cent of potassium nitrate to laboratory media completely inhibits the growth of the bacillus. As the pickling solutions always contain considerably more than these percentages of sodium chlorid and potassium nitrate, it would be impossible for the bacillus to multiply in the pickles. Additional laboratory experiments demonstrated, however, that the bacillus or its spores may remain alive in the curing pickles for at least thirty days, and it seemed possible that the curing pickles might become contaminated at times with the bacilli, and that the bacilli, although incapable of multiplying in the pickles, might find their way into the bodies of the hams in the pickling fluids. In order to throw some light upon this point, the following experiment was carried out:

EXPERIMENT TO SHOW WHETHER INFECTION TAKES PLACE FROM THE CURING PICKLE.

In this experiment two tierces were put down, each containing 20 hams. The hams weighed from 14 to 16 pounds and had received the usual 48-hour chilling. The pickling solutions employed were the regular curing pickles of the establishment at which the experiment was carried out. The curing pickle in one tierce was inoculated with 400 cubic centimeters of a culture suspension of the bacillus, prepared in the same manner as that used for the injection of the hams in tierce 2 in Experiment II. A microscopic preparation made from a small drop of the culture suspension before adding it to the pickle showed the bacilli in large numbers, and in the 400 cubic centimeters of the suspension there were millions of the bacteria. The curing pickle in the other tierce was left untreated, the hams in this tierce serving as a check. The tierces used in this experiment, as in all of the experiments, were thoroughly cleaned with boiling water before the hams were placed in them. The experiment was conducted in a pickling room which was held at 33° to 36° F., and the tierces were rolled three times during the cure. The details of the experiment are as follows:

Tierce 1.—Contained 20 hams, half of which were pumped in both body and shank and half in the shank only. As soon as they were pumped the hams were packed in the tierce. Sufficient curing pickle to fill the tierce was then measured out in a clean barrel and to it was added the culture suspension. The culture was thoroughly mixed with the pickle and the latter was then run into the tierce containing the hams.

Result: When tested at the end of the cure, two of the hams which had been pumped in the shank only showed slight souring in the body. The rest of the hams were sweet.

Tierce 2.—Contained 20 hams which were pumped in the same manner as those in tierce 1. The curing pickle was the same as that used for tierce 1, but without the addition of culture. This tierce was put down as a check on tierce 1, the hams being cured under exactly the same conditions, but without the addition of culture to the curing pickle.

Result: One of the hams which was pumped in the shank only developed slight souring in the body. The remainder of the hams were sweet.

Comparing tierce 1, which contained the inoculated pickle, with tierce 2, the check tierce which contained uninoculated pickle, we find there was practically no difference in the final result. In tierce 1 two of the hams developed slight souring, while in tierce 2 one of the hams became slightly sour. All of these hams had been pumped in the shank only. The fact that one of the hams in the check tierce developed slight souring was undoubtedly due to bacterial contamination in pumping or in the handling which the hams underwent prior to pickling, and the slight souring of the two hams in tierce 1 must also be attributed to the same cause or causes, for had the souring in these last hams resulted from the penetration of the bacteria from the pickling solution a higher percentage should have become sour. Furthermore, if the souring of the two hams in tierce 1 had resulted from the penetration of the bacteria from the curing pickle, the souring should have been general throughout the bodies of these hams, whereas the souring was only evident around the bone and was slight in degree.

From this experiment the conclusion would seem justified that the bacillus which causes ham souring does not usually find its way into the bodies of the hams from the curing pickle, although it would be going too far, perhaps, to say that infection never takes place from the curing pickle. The experiment, however, indicates clearly that the curing pickles are certainly not the main channel through which the hams become infected. In referring to the curing pickles, it should be understood that we refer here to the pickling solutions in which the hams are immersed, and not to the pumping pickles. The possibility of infection through the pumping pickle will be discussed later.

POSSIBLE INFECTION THROUGH MANIPULATION OR HANDLING.

There are at least three possible ways in which hams may become infected from the handling which they receive in preparation for, or during the process of curing, viz: From the thermometers used in taking the inside temperatures of the hams, from the pumping needles, and from the billhooks used in lifting the hams.

INFECTION FROM HAM THERMOMETERS.

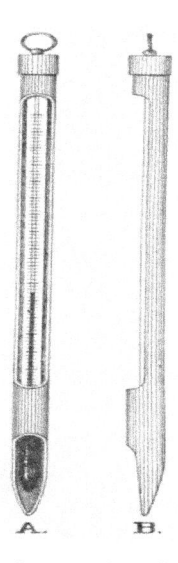

FIG. 4.—**Diagrammatic views showing construction of ham thermometer. A, front view, showing open space between metal point and mercury bulb, which becomes filled with particles of meat, grease, and dirt; B, side view.**

The packing-house method of taking the temperatures of hams by means of a pointed, metal-capped thermometer which is thrust deep into the bodies of the hams has already been referred to, but deserves to be described somewhat more in detail, as it will be at once apparent that this manipulation furnishes a ready means whereby hams may become infected with putrefactive bacteria. The construction of a ham thermometer is shown in figure 4.

The instrument consists of a glass thermometer inclosed in a metal case, the front portions of the case being cut away so as to expose the scale above and the mercury bulb below. As was explained before, the thermometer is thrust deep into the body of the ham so that the pointed end containing the mercury bulb rests beside or a little behind the upper portion of the femur, the bone being used as a guide in introducing the thermometer.

Ham temperatures are taken at three stages in the preparation for cure—(1) on the hanging floor, just before the hams go to the chill rooms, in order to determine the amount of heat lost prior to chilling; (2) on leaving the chill rooms, in order to determine the thoroughness of the chill;(3) on the packing floor, just before the hams are placed in pickle, as a further check on the thoroughness of the chilling.

In taking the temperatures of hams which have been chilled—and most of the temperatures are taken subsequent to chilling—it is customary for the packing-house attendant who has this matter in charge to warm the thermometer by holding the pointed or bulb end in his hand, so as to force the mercury column up to about 60° F., or well above the temperature of hams. The thermometer is then thrust into the ham and allowed to remain for several minutes, by which time the mercury column will have fallen to the temperature of the ham. The thermometer is then slowly withdrawn so as to expose the top of the mercury column, and an accurate reading is thus obtained of the inside temperature of the ham. The thermometer is warmed by the hand before each ham is tested, and this undoubtedly insures more accurate readings than would result were the thermometer removed from one ham and plunged immediately into another, but the procedure is open to certain objections, for the open space between the metal point of the thermometer and the mercury bulb soon becomes filled with particles of meat and with grease and dirt from the attendant's hands, and it is at once apparent that a thermometer in this condition would furnish a ready means whereby extraneous matter might be introduced into the bodies of the hams. In other words, a contaminated thermometer would furnish an excellent means whereby hams could be inoculated with putrefactive bacteria.

In order to determine whether hams actually become inoculated in this manner, the following experiment was carried out:

EXPERIMENT TO SHOW WHETHER HAMS BECOME INFECTED FROM HAM THERMOMETERS.

This experiment was designed to show (1) whether the usual packing-house method of taking ham temperatures was apt to induce souring in the hams thus tested, and (2) whether souring would result in hams which were tested with a thermometer purposely contaminated with the bacillus isolated from sour hams.

The experiment was carried out as follows: Thirty hog carcasses were selected as they entered the hanging floor from the killing floor. They had been cleaned, eviscerated, and split, and were of the same average weight and of sufficient size to yield hams weighing from 12 to 14 pounds. They were divided into three lots of 10 each and were allowed to remain on the hanging floor for two hours, after which they were given the usual 48-hour chilling.

Lot 1.—The hams in this lot were tested with an ordinary ham thermometer as they entered the hanging floor, as they left the hanging floor, and as they left the coolers. The thermometer used was borrowed from one of the plant attendants and was used in the condition in which it was received from him; that is, it was not cleaned or disinfected prior to use.

Lot 2.—The hams in this lot were tested as they entered the hanging floor with a thermometer which had been previously cleaned and disinfected and then dipped in a culture suspension of the meat-souring bacillus which was isolated from sour hams. The thermometer was dipped in the culture suspension before each ham was tested. No further temperatures were taken of these hams. The thermometer was carefully cleaned and disinfected before it was returned to the attendant from whom it was borrowed.

Lot 3.—The hams in this lot were not tested at all, and were intended as checks on the cure.

The three lots of carcasses were carefully tagged and were chilled in a special cooler to themselves. Upon leaving the cooler the hams were cut from the carcasses and trimmed. The three lots of hams were then cured in separate tierces. All of the hams were subjected to exactly the same cure.

The pickles used were the regular pumping and regular curing pickles of the establishment at which the experiment was carried out.

The hams in lot 3 were pumped first and those in lot 1 were pumped next. The needle was then removed and a fresh, clean needle was used for lot 2.

This was done in order to prevent the possibility of carrying over bacteria from one lot of hams to another on the pumping needle. The tierces were thoroughly cleaned with boiling water before being used. The curing was carried out in a pickling cellar which was held at 33° to 36° F., the temperature never rising above the latter figure. The tierces were rolled three times during the curing. The details and results were as follows:

Tierce 1.—Contained 20 hams, half of which were pumped in both body and shank and half in the shank only. These hams were taken from the carcasses in lot 1 and had been tested several times with a ham thermometer, as already described.

Result: At the end of the cure it was found that of the 10 hams which were pumped in the shank, 5 showed well-marked souring in the body, while of the 10 hams which were pumped in both body and shank, 2 showed slight souring in the body.

Tierce 2.—Contained 20 hams, which were pumped in the same manner as those in tierce 1. These hams were taken from the carcasses in lot 2 and had been tested once with a thermometer which was dipped in a culture suspension of the bacillus isolated from sour hams.

Result: The 10 hams which were pumped in the shank only all became sour. When they were tried out at the end of the cure, they showed pronounced souring throughout the entire body and were classed as "stinkers" by the meat inspector who examined them. The souring extended through to the bone marrow of the femur in all of these hams. Of the 10 hams which were pumped in both body and shank, 7 showed well-marked souring in the body, but not as pronounced as in those pumped in the shank only; in five of these hams the souring extended through to the bone marrow of the femur, while in 2 the bone marrow remained sweet.

Tierce 3.—Contained 20 hams, which were pumped in the same manner as those in the two preceding tierces. These hams were not tested with a thermometer, and were put down as a check on the cure. They were pumped with the same pumping pickle, subjected to the same cure, and held under precisely the same conditions as the hams in the two preceding tierces.

Result: When tested at the end of the cure, all of these hams were found to be perfectly sound and sweet.

Results of experiment to show whether hams become infected from ham thermometers.

No. of tierce.	Number of hams.	Average weight of hams. Pounds.	How pumped.	Treatment.	Condition at end of cure. Number of sour hams.	Percentage of sour hams.
1	20	12-14	10 in shank	Tested in several stages in preparation for cure which had not been cleaned.	5	50
			10 in body and shank	do	2	20
2	20	14-16	10 in shank	Tested once with ham thermometer dipped in culture suspension of anaerobic bacillus isolated from sour hams.	10	100
			10 in body and shank	do	7	70
3	20	14-16	10 in shank	Not tested with thermometer.	0	0
			10 in body and shank	do	0	0

Several hams from each tierce were examined bacteriologically. cultures being taken from the meat near the bone and from the bone marrow of the femur.

In the sour hams from tierce 1 cultures taken from the meat near the bone showed the same anaerobic bacillus noted in other sour hams (i. e., the same bacillus which caused souring in Experiments I and II), but these cultures were contaminated with other bacteria which were probably introduced on the thermometer along with the ham-souring bacillus. None of the contaminating bacteria were capable, however, of producing a sour-meat odor when grown on the egg-pork medium. Pure cultures of the ham-souring bacillus were obtained from the bone marrow of some of these hams, showing that this bacillus had penetrated through to the bone marrow while the other bacteria had not.

From the sour hams in tierce 2 the ham-souring bacillus was recovered readily, and often in pure culture, from the hams which had been pumped in the shank only, whereas it was usually contaminated with pickle bacteria in the hams which had been pumped in both body and shank.

In the case of the sound hams in tierce 3, cultures taken from the meat near the bone and from the bone marrow of the femur were negative in the hams which had been pumped in the shank only, while cultures taken from corresponding points in the hams pumped in both body and shank showed ordinary pickle bacteria, which had evidently been introduced into the bodies of these hams in the pumping pickles. None of these hams exhibited the slightest sour odor.

Summary of experiment.—In this experiment 20 hams (tierce 1) were tested with an ordinary ham thermometer in the usual packing-house manner. Half of these hams were subjected to the mild cure and half were given the regular cure, with the result that 50 per cent of those receiving the mild cure and 20 per cent of those receiving the regular cure became sour.

A second lot of 20 hams (tierce 2) were tested with a thermometer which had been purposely contaminated with a culture suspension of the ham-souring bacillus. These hams were cured in the same manner as the first lot, with the result that all of those receiving the mild cure and 70 per cent of those receiving the regular cure became sour.

A third lot of 20 hams (tierce 3) which had not been tested at all were cured in the same manner as the two preceding lots, as a check on the cure. All of these hams were sweet at the end of the cure.

Inasmuch as the three lots of hams were cured under precisely the same conditions and were handled in the same manner prior to pickling, the only difference being that the hams in tierces 1 and 2 were tested with the ham thermometer while those in tierce 3 were not, we must conclude that the souring of the hams in tierces 1 and 2 resulted from the testing which these hams received. In the case of tierce 1 the hams became infected from a thermometer which, in the ordinary routine use of the packing house, had become accidentally contaminated with the ham-souring bacillus. In the case of tierce 2 the hams became infected from a thermometer which had been artificially contaminated with the bacillus. The high percentage of sours in this last lot is due to the fact that these hams were heavily infected with the ham-souring bacillus, for owing to the construction of the ham thermometer many thousands of the bacilli were unquestionably introduced into each ham on the point of the thermometer. In the ordinary routine of ham testing, where hams become infected from foreign matter introduced on the thermometer, the percentage of souring, as shown in tierce 1, would be less, for it is not to be supposed that ham thermometers are always contaminated with the ham-souring bacillus, but that they only become so at times, and that probably only a few of the bacilli are then introduced.

This experiment, we think, proves conclusively (1) that the ham-souring bacillus may be introduced into the bodies of hams on the thermometers used in testing the hams, and (2) that the packing-house method of taking ham temperatures by means of a thermometer which is thrust deep into the bodies of the hams may cause souring in the hams thus tested.

As a further proof that hams may become contaminated in this manner, a series of cultures were made from scrapings taken from ham thermometers. The scrapings consisted of the accumulations of bits of meat, grease, and dirt that collect on the thermometers, and were taken from the thermometers while the latter were in ordinary routine use in the packing house. In a series of six cultures which were made from such scrapings at different times, the same bacillus which was isolated from sour hams and shown to cause meat souring was found three times. In other words, the ham-souring bacillus was present in 50 per cent of the cultures made from thermometer scrapings, and many hams undoubtedly become infected from the thermometers. Souring would be almost certain to result in mild-cure hams if these hams were tested with a thermometer which had become accidentally contaminated with the ham-souring bacillus, as the bacillus would have time to develop within the bodies of the hams before being inhibited by the curing pickle, which penetrates slowly into the bodies of these hams. In the case of regular cure hams—that is, hams which are pumped in both body and shank—souring would be much less apt to occur after the use of a contaminated thermometer, as these hams are more or less saturated with a strong pumping pickle at the beginning of the cure, which would tend to inhibit the growth of any bacteria that might be introduced on the thermometers.

The fact that souring may result in hams from the use of a contaminated thermometer would explain the occurrence of several sours in one vat, for in testing hams just before they go into cure several hams are usually tested in succession, and these would in all likelihood go into the same vat. Supposing the thermometer to have been contaminated with the ham-souring bacillus at the time these hams were tested, this would explain a fact which has been often noted, namely, the occurrence of several sours in one vat while other vats containing the same run of hams show no sours.

If souring resulted in all of the hams which are subjected to a thermometer test in the daily routine of the packing house, this manipulation alone might account for nearly all of the sours which occur, but the experiment which has been just described shows that all of these hams do not become sour. In tierce 1, where each ham was subjected to three thermometer tests at different times, souring resulted in 35 percent (this includes both mild and regular cure) of the hams thus tested, and in actual practice the percentage of sours in hams which have been subjected to the thermometer test would probably be somewhat less. Quite a large percentage of sour hams are thus left unaccounted for by the thermometer test, and we believe that these are chiefly the result of contamination carried in on the pumping needles or in the pumping pickles.

INFECTION FROM PUMPING NEEDLES.

In view of the results obtained in the last experiment, in which it was shown that hams may become infected from the use of ham thermometers, it seemed not improbable that hams might also become infected from the pumping needles, which, like the thermometers, are thrust deep into the bodies of the hams beside the bone. In order to throw some light upon this point, cultures were taken from the grease and dirt that accumulate on the shields at the bases of the pumping needles, as such material must undoubtedly be carried into the hams at times on the needles. The ham-souring bacillus was found several times in these cultures, and hence it is fair to infer that hams may also become infected at times from the pumping needles, just as they become infected from the thermometers. Bits of contaminated meat and grease and particles of dirt carried in on the pumping needles would be forced out into the hams by the pumping pickle, which passes out through small openings or fenestræ in the needles, and this probably affords one explanation as to why so many more body sours occur in the mild-cure hams. In the mild-cure hams, which are pumped in the shank only, the pumping needle is introduced near the femorotibial articulation, and the shank is saturated at the start with a strong brine solution, while the body of the ham is not. If the ham-souring bacillus were carried into these hams on the pumping needle, the growth of the bacillus in the shank would be inhibited by the strong brine solution with which the shank is saturated, but there would be nothing to prevent the bacillus from growing upward into the body of the ham, which has not been pumped and is free from pickle. This would also explain the fact that the souring often starts at the knee joint and extends upward into the body of the ham. In the case of the regular cure hams, where the ham is pumped in both body and shank, the entire ham is more or less saturated at the start with the strong brine of the pumping pickle, which tends to inhibit the growth of the ham-souring bacillus even if this bacillus should find its way into these hams on the pumping needles. It is in the mild-cure or partly pumped hams, where the body of the ham is left unpumped, that the ham-souring bacillus finds its best opportunity for development, and the greater proportion of the sours that occur in the packing house are found in these hams.

As regards the possibility of infection from the pumping pickle itself, it does not seem probable that this would often occur, for the pumping and curing pickles are always prepared on an upper floor of the pickling houses and are delivered to the pickle cellars in closed pipes, so the chances for the accidental contamination of these solutions from floating dust or dirt would not be great. Furthermore, the strong brine of the pumping pickle would completely inhibit the growth of the ham-souring bacillus, and the bacillus would be incapable of multiplying, even if it found its way into the pickle. On the other hand, laboratory experiments show that the bacillus or its spores may remain alive for a considerable length of time in the pumping pickle, so the possibility of infection from this source can not be overlooked.

INFECTION FROM BILLHOOKS.

After the hams are cut from the carcasses they are handled entirely by means of billhooks. In handling the hams the hooks are inserted beneath the skin of the shank at a point just above the tibio-femoral articulation. The hooks should be inserted in the connective tissue beneath the skin and should not penetrate the muscular tissue to any depth. When the hams lie in the right position, with the butt or large portion away from and the shank toward the operator, it is an easy matter to pick them up in the proper manner; but when they lie at different angles and are being rapidly handled it is almost impossible to prevent the hook from penetrating the muscular tissues, and if the hook should penetrate to the bone it might carry in foreign matter contaminated with the meat-souring bacillus. It is not probable that many hams become contaminated in this way, as the men who handle the hams are very skillful in manipulating their hooks; but the possibility that hams may become contaminated in this manner should not be entirely overlooked.

BIOLOGICAL AND MORPHOLOGICAL CHARACTERISTICS OF THE HAM-SOURING BACILLUS.

CONDITIONS FAVORABLE TO GROWTH.

The most favorable medium for the growth of the organism was found to be the modified egg-meat mixture of Rettger, which has been previously described. In this medium the organism develops rapidly at a temperature of 20° to 25° C., giving rise to the characteristic sour-meat odor. Like the bacillus described by Klein, it also grows readily on pork-agar and pork-bouillon containing glucose, but differs from Klein's bacillus in that it will grow, though less luxuriantly, on ordinary nutrient media—agar, gelatin, and bouillon—without the addition of glucose.

The optimum temperature for growth is 20° to 25° C. The organism does not grow at incubator temperature (37.5° C.). At ice-box temperature (8° to 10° C.) it develops readily, although the growth is less rapid than at 20° to 25° C. That the organism will develop at even lower temperatures was shown in the inoculation experiments with hams, where it developed and multiplied extensively in the bodies of the hams at the temperature of the pickling cellars, which are held usually at 34° to 36° F.(1° to 2° C.).

The organism develops best in a neutral or slightly alkaline medium.

GROWTH ON DIFFERENT CULTURE MEDIA.

Growth on egg-pork medium.—At a temperature of 20° to 25° C. the cultures show a slight but distinct sour odor in from two to three days. This odor, as before stated, closely resembles the odor of a sour ham. Egg-pork cultures from three to five days old were given to a trained meat inspector, who knew nothing whatever as to the contents of the tubes, and he was asked to describe the odor; he described it as that of a sour ham.

At one week the albumins of the medium are gelatinized or partly coagulated and the odor is more pronounced. At ten days the albumins are completely coagulated except at the surface, where there is no apparent growth; the odor is more putrefactive in nature, and the reaction of the medium is slightly acid. At three weeks the coagulated albumin splits up into fragments and appears to undergo a slow digestion, gas bubbles form in the lower portion of the culture, and the odor becomes distinctly putrefactive in character. The slow digestion of the albumin is probably due to a proteolytic enzyme elaborated by the bacillus.

At the end of a week a dark zone usually appears at the surface of the coagulated albumin and gradually darkens until it becomes almost black. This zone is probably due to a pigment elaborated by the bacillus.

At ice-box temperature (8° to 10° C.) the same changes and the same odor were noted, but were somewhat slower in developing.

Glucose-pork-agar.—This medium was prepared from pork in the same manner as beef-agar, and contained 1 per cent of glucose. The organism grows readily on this medium and may be conveniently cultivated in deep stab cultures. The medium was always thoroughly boiled and then rapidly cooled in order to expel the inclosed air. The growth of the organism was found to vary considerably with the reaction.

When the reaction was +1.5, deep stab cultures at three days (20° to 25° C.) showed a well-marked arborescent growth, appearing as delicate filaments extending outward from the line of stab. The growth stopped within one-fourth or one-half inch of the surface of the agar on account of the presence of oxygen in the upper part of the culture medium. As the growth extended toward the walls of the test tube the agar became clouded, and there were sometimes gas bubbles in the depth of the agar, but the gas formation was not extensive.

When the reaction of the agar is neutral or slightly alkaline, extensive gas formation occurs and the agar is often much broken up.

The cultures developed a disagreeable, somewhat putrefactive odor, but did not give the characteristic sour-ham odor obtained from the egg-pork cultures.

The organism was also grown on anaerobic agar plates by Zinsser's method, which is said to give absolutely anaerobic conditions. The colonies on agar have a cottony or woolly appearance at first, and spread slowly, with slightly irregular margins.

In glucose-pork-agar to which azolitmin was added the azolitmin in the lower portion of deep stab cultures was completely decolorized in five days at room temperature (20° to 25° C).

In glucose-pork-agar containing neutral red the red color in the lower portion of the tube was changed to yellow with the development of fluorescence.

Neutral gelatin.—Tubes of ordinary neutral gelatin without the addition of glucose were inoculated and held at ice-box temperature (8° to 10° C). At five days a delicate white growth appeared along the line of stab in the lower portion of the tube. At seven days the growth showed fine radial striæ, presenting an arborescent or tree-like appearance, and extended halfway from the line of stab to the walls of the test tube. At two weeks the growth had caused a delicate clouding of the medium in the lower portion of the tube. At three weeks the gelatin in the lower portion of the tube had become liquefied and the growth had settled to the bottom as a white precipitate.

In gelatin containing glucose, gas bubbles are formed in the depth of the medium through the splitting up of the glucose, and the characteristic arborescent growth is obscured.

Glucose-pork-bouillon.—This medium was prepared from pork instead of beef and contained 1 per cent of glucose. The best results were obtained when the reaction of the medium was neutral or slightly alkaline.

Culture tubes, which had been previously boiled to expel the contained air and then inoculated, were held in a Novy jar, in an atmosphere of hydrogen at a temperature of 20° to 25° C. At three days the tubes showed well-marked clouding. At one week the growth appeared as a heavy, white, flocculent, cottony precipitate in the bottom of the tubes with a slight flocculent precipitate above. When the culture was removed from the jar and shaken, the heavy, flocculent precipitate at the bottom of the tube broke up without much difficulty, giving rise to a heavy uniform clouding with some small floating masses, which soon settled to the bottom. On shaking the tube some evolution of gas in the form of very fine bubbles was noticed.

In Smith fermentation tubes containing neutral glucose-pork-bouillon the closed arm of the tube shows well-marked clouding with gas formation at three days at room temperature (20° to 25° C). The growth has a tufted, cottony appearance, and there are many filaments and threads. The growth settles to the bottom of the closed arm as a cottony, white precipitate (see Pl. IV). The organism splits the glucose vigorously, and at 10 days the tubes show from 40 to 50 per cent of gas. The bouillon in the open arm of the tube remains unclouded. The maximum gas production at room temperature is reached in from 10 to 14 days, by which time the growth in the closed arm has completely settled into the bend of the tube, leaving the bouillon in the closed arm clear. The gas formula, as determined by Smith's method, was $H/CO_2 = 5/1$. The reaction of the bouillon becomes acid to phenolphthalein.

The organism will grow on ordinary neutral bouillon without the addition of glucose, and in Smith tubes containing this medium a small amount of gas was formed, due to the splitting of the muscle sugar.

The bacillus also grows in a sugar-free broth—that is, a broth free from muscle sugar—and from cultures grown in this medium a well-marked indol test was obtained.

Litmus-milk.—The organism was grown in litmus-milk in Smith fermentation tubes at 20° to 25° C. At seven days the litmus in the lower portion of the closed arm had assumed a brownish-buff color. At two weeks the litmus in the closed arm had been reduced to a brownish-buff color except at the top of the tube, where a pale, bluish tinge remained, and the litmus in the open arm showed very slight reddening as compared with a check tube. At three weeks the litmus in the closed arm was entirely reduced to a light, brownish-buff color, and the litmus in the open arm showed a slight but distinct reddening as compared with the check. The reddening of the litmus in the open arm was evidently due to the transfusion of acids formed by the growth of the bacillus in the closed arm. After several weeks the milk is slowly peptonized, probably as a result of enzyme action.

MORPHOLOGY.

The organism is a large bacillus having an average size of 4 to 8 μ in length by 0.5 to 0.7 μ in thickness, but there are many longer forms measuring from 10 to 20 μ in length. It develops in long, irregular chains or filaments, which at times show a slightly spiral form.

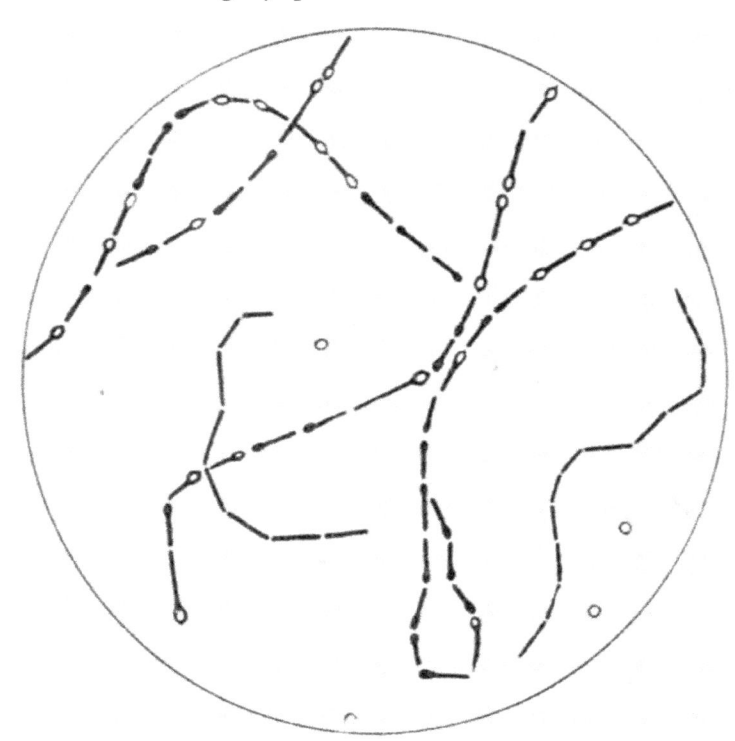

FIG. 5.—Ham-souring bacillus (*Bacillus putrefaciens*) grown on egg-pork medium, showing tendency to form chains. Partly developed and fully developed spores are shown at ends of rods; also free spores. (Pen-and-ink drawing made with camera lucida from preparation stained by Gram's method.× 640.)

The individual organisms show at times a widely open, slightly spiral form, which was more apparent in hanging-drop preparations made from bouillon cultures, where the organisms had been comparatively undisturbed. This appearance was also noted at times in the stained sections of soured muscular tissue, where the organisms were stained in place. The organism possesses no motility. It stains with the ordinary aniline dyes and by Gram's method.

SPORE FORMATION.

The organism develops large, terminal spores, which are at first oval, but when fully developed are perfectly round and measure from 1.5 to 2 μ in diameter.

Spores develop rapidly in the egg-pork medium at 20° to 25° C., fully developed spores being noted in from five to seven days. At ice-box temperature (8° to 10° C.) partly developed spores were noted in the egg-pork medium at 10 days and fully developed spores at 2 weeks.

Occasional spores were noted in old agar and gelatin cultures, but abundant spore formation was seen only in the egg-pork medium. No spores were noted in bouillon cultures, even at 10 weeks.

RESISTANCE TO HEAT AND CHEMICAL AGENTS.

In its vegetative form the bacillus is killed at 55° C. in 10 minutes. The spores survive a temperature of 80° C. for 20 minutes, but are killed at 100° C. in 10 minutes.

When sodium chlorid and potassium nitrate were added to glucose-pork broth in varying amounts, it was found that 3 per cent of sodium chlorid or 3 per cent of potassium nitrate was sufficient to inhibit completely the growth of the bacillus at room temperature (20° to 25° C.).

While the growth of the bacillus was inhibited by sodium chlorid and potassium nitrate as just stated, it was found that very much stronger solutions of the two salts failed to destroy the bacillus. Thus it was found that the bacillus or its spores retained their vitality after an exposure of 30 days in a solution containing 23 per cent of sodium chlorid and 6 per cent of potassium nitrate.

GAS PRODUCTION.

The organism splits glucose, but not lactose or saccharose. That it possesses the power of splitting muscle sugar was shown by the formation of gas in Smith fermentation tubes containing ordinary neutral bouillon without the addition of any sugar.

The formation of gas in glucose bouillon varies considerably with the reaction of the medium. The largest amount of gas was formed when the broth was neutral or slightly alkaline. When the reaction of the broth was distinctly acid or distinctly alkaline the amount of gas was diminished. The gas which is formed in bouillon cultures consists chiefly of hydrogen and carbon dioxide. In order to collect a sufficient amount of the gas for analysis, two large fermentation tubes capable of holding 150 cubic centimeters each were constructed. These tubes were filled with pork-bouillon and inoculated with the bacillus. After 20 days at room temperature (20° to 25° C.) the gas was collected and the carbon dioxide and hydrogen determined, with the following result:

	Cubic centimeters.
Total amount of gas collected	37.7
Carbon dioxide, by absorption with NaOH	6.2
Hydrogen, by difference	31.5

This analysis gives an approximate gas formula of $H/CO_2 = 5/1$, which agrees with the gas formula as determined in the small fermentation tubes by Smith's method.

In hams which had undergone spontaneous souring and in hams which had been artificially soured by inoculation, hydrogen-sulphid was often noted when the sour portions of the meat were tested with lead-acetate paper, but no distinct odor of the gas could be obtained. Hydrogen sulphid was also noted in egg-pork cultures of the bacillus.

ACID PRODUCTION.

In glucose-bouillon, butyric and lactic acids are formed and the reaction of the medium becomes distinctly acid. Butyric and lactic acids were also noted in the egg-pork cultures.

A series of Smith fermentation tubes containing 10 c. c. each of glucose-pork broth medium was inoculated with the bacillus and held at room temperature (20° to 25° C.). These cultures were titrated against [N/40]NaOH, with phenolphthalein as an indicator at intervals of two days up to nineteen days, and then at two-week intervals up to sixty-one days. Three of the cultures were titrated each time so as to give a fair average of the acidity of the cultures, and an uninoculated check tube was also titrated each time to see if there was any change in the reaction of the medium. The results of the titrations are shown in the following table:

Acidity determinations in glucose-pork broth cultures.

Age of culture (days).	Culture A.	Culture B.	Culture C.	Average.	Medium.	Acidity of culture.
						Per cent.
2	0.038	0.030	0.040	0.036	0.009	0.027
4	.105	.100	.102	.102	.009	.093
6	.106	.110	.109	.108	.009	.099
8	.124	.115	.117	.119	.009	.110
10	.128	.130	.126	.128	.009	.119
12	.129	.120	.129	.126	.009	.117
19	.126	.125	.125	.125	.009	.116
33	.125	.123	.125	.124	.009	.115
47	.122	.120	.121	.121	.009	.112
61	.121	.116	.119	.118	.009	.109

From the above table it will be seen that the maximum acidity was reached at ten days, after which there was a gradual reduction in the acidity, due probably to the formation of ammonia compounds.

PATHOGENIC PROPERTIES.

Rabbits, guinea pigs, and white mice were inoculated and fed with cultures of the bacillus without effect, from which it would appear that the bacillus possesses no pathogenic or disease-producing properties.

NATURE OF THE BACILLUS.

The bacillus is essentially a saprogenic bacterium with zymogenic properties. A preliminary study of the chemical changes which take place in sour hams shows that these changes are of a putrefactive nature. Hams which had undergone spontaneous souring were compared with hams which had been artificially soured by inoculation, and the chemical changes were found to be identical. A chemical study was also made of the changes taking place in egg-pork cultures of the bacillus at different stages of growth, and these changes were found to be of a putrefactive nature and similar in character to the changes which occur in sour hams. Among the putrefactive products formed by the growth of the bacillus in the egg-pork medium were indol, skatol, volatile fatty acids, skatol-carbonic acid, and hydrogen sulphid.[4]

[4] The tests for the putrefactive products formed by the growth of the bacillus in the egg-pork medium were made by P. Castleman, of the Biochemic Division, who also determined the percentage composition of the gas formed by the growth of the bacillus in the glucose-pork-bouillon medium.

BUL. 132, BUREAU OF ANIMAL INDUSTRY, U. S. DEPT. OF AGRICULTURE. PLATE IV.

GLUCOSE BOUILLON CULTURE IN SMITH FERMENTATION TUBE AT FOUR DAYS. CULTURE GROWN AT ROOM TEMPERATURE (20° TO 25°

C.). GROWTH CONFINED ENTIRELY TO CLOSED ARM, WITH GAS COLLECTING AT TOP.

A more extended study is now being carried on in the Biochemic Division of the chemical changes which take place in hams during the process of souring, together with a further study of the chemical changes which result from the growth of the bacillus in the egg-pork medium. The results of this investigation will be given in a later paper.

The bacillus described in this paper belongs to the class of putrefactive anaerobes, which are widely distributed in nature in dust, soil, and excrementitious matters. This group of bacteria contains both pathogenic and nonpathogenic forms. The former have received considerable attention, but the latter have never been thoroughly cleared up. The bacillus isolated from sour hams belongs in the latter category, being possessed of no pathogenic or disease-producing properties. It occurs in the dust and dirt of the packing house and finds its way into the hams in the various manipulations to which the hams are subjected.

The bacillus described in this paper does not seem to correspond with any forms heretofore described. It differs from Klei bacillus (*Bacillus fœdans*) in the following important particulars:(1) It forms large terminal spores, whereas Klein's bacillus formed no spores;(2) it will grow at a temperature of 34° F., while Klei bacillus did not grow below 50° F.;(3) it produces an acid reaction in culture media, while Klei bacillus gave a distinctly alkaline reaction;(4) it will grow on the ordinary nutrient media—gelatin, agar, and broth—without the addition of glucose, while Klein's bacillus did not;(5) it peptonizes the casein in milk, whereas Klein's bacillus had no action on milk; (6) it liquefies gelatin more rapidly, causing complete liquefaction after three weeks at 8° to 10° C., whereas Klein's bacillus caused only partial liquefaction after eight weeks at 20° C.;(7) it can be conveyed

from turbid broth cultures to new culture material by means of the platinum loop, whereas Klein's bacillus could not be thus conveyed.

For the bacillus described in the present paper the following name is proposed:*Bacillus putrefaciens*.

PREVENTION OF HAM SOURING.

As it has been shown that souring in hams results from the growth of a bacterium which is introduced into the bodies of the hams in the various manipulations which the hams undergo, the only way to eliminate souring in hams, as they are cured in the larger packing establishments, would be to cure the hams under aseptic or sterile conditions, which would, of course, be a physical impossibility.

While it will probably be impossible, therefore, to eliminate souring entirely under the methods of ham curing which are at present employed in the larger packing establishments, much can undoubtedly be done toward reducing the percentage of sours. In the matter of taking ham temperatures, for instance, if the thermometers used were thoroughly cleaned and disinfected and the surfaces of the hams seared at the point where the thermometer is introduced, infection from this source could be entirely prevented; or it might be possible so to regulate the temperature of the chill rooms that the taking of ham temperatures could be discontinued.

The elimination of the souring that results from the introduction of foreign matter on the pumping needles could be effected in two ways only,(1) by not pumping the hams at all, or (2) by pumping them under sterile or aseptic conditions. As has been stated before, some of the smaller packing establishments cure their hams without pumping, and in these establishments the percentage of sours runs very low. When hams are cured without pumping, however, the period of curing has to be materially lengthened in order to give the curing pickles sufficient time to penetrate thoroughly, and this is what the larger plants wish to avoid because of the greater space and greater number of vats which would be necessitated. The object of pumping in the larger plants, where the number of hams handled daily runs into the thousands, is to hasten the cure and thus prevent the accumulation of a great number of hams at one time. It is doubtful, therefore, whether the larger packing houses could conveniently discontinue pumping.

To pump the hams under aseptic conditions would necessitate a technique far too elaborate for routine use in the packing house; in fact, anything like complete asepsis would be out of the question. Certain measures might be adopted, however, that would tend to prevent the possible introduction of ham-souring bacilli in the process of pumping. It would undoubtedly be safer, for instance, to boil the pumping pickle before use, and the chances of carrying in contaminated foreign matter on the pumping needles could be lessened by sterilizing the pumps and needles with boiling water and by frequently dipping the needles, while in use, in boiling water. If the hams were sprayed with clean water just prior to pumping, there would be less likelihood of carrying in foreign matter on the needles. The danger of introducing contaminated foreign matter on the needles might be further obviated by searing the surfaces of the hams at the points where the needles are introduced; but such a procedure would be hardly practicable in the larger packing houses, where the great number of hams cured necessitates rapid handling.

While the danger of possible contamination in pumping, through the introduction of contaminated foreign matter on the pumping needles, can not well be avoided, this danger is partly counterbalanced by the inhibitory action of the pumping pickle, which is strikingly shown in the experiments which have been described. In these experiments, 100 hams received large doses of the ham-souring bacillus, half of these hams being subjected to the mild cure and half to the regular cure, with the following result: In the case of the mild-cure hams, which were pumped in the shank only, the percentage of sours was practically 100 per cent, every ham with possibly one exception becoming sour; whereas in the regular-cure hams, which were pumped in both body and shank, only 58 per cent of the hams became sour. In other words, the additional pumping which the regular-cure hams received served to prevent souring in 42 per cent of these hams. In these experiments the number of bacteria introduced into the hams was very great, thousands and even millions of the bacilli being introduced into each ham, whereas in the routine of the packing house it is not likely that more than a few of the bacilli are ever introduced at one time on the thermometers and pumping needles. In view of these results it is safe to say that in the larger packing houses, where pumping seems to be necessary, the number of sours could be reduced fully 50 per cent if all hams were pumped in the body as well as in the shank.

At present the usual procedure is to pump all hams, both mild and regular cure, with the same pumping pickle, the mild-cure hams being pumped in the shank only and the regular-cure hams at two additional points in the body. The experiments quoted above show that the additional pumping which the regular-cure hams receive undoubtedly tends to prevent the development of souring in these hams, and this result is unquestionably due to the inhibitory action of the salts contained in the pumping pickle, as it was found by laboratory experiment that the addition of 3 per cent of sodium chlorid to culture media is sufficient to inhibit the growth of the ham-souring bacillus. The pumping pickles consist of strong brine solutions and always contain considerably more than 3 per cent of sodium chlorid. If, therefore, the pumping of regular-cure hams were made more thorough than at present, and all of the deeper portions of the ham were thoroughly saturated with the strong brine solution, souring could be largely eliminated, if not entirely prevented, in these hams, as an unfavorable medium or soil would thus be created in which the ham-souring bacillus could not develop. The ham-souring bacillus is able to develop within the bodies of the regular-cure hams because the pumping of these hams is not always thorough and there are certain areas in the inner or deeper portions of the hams in which the tissues are not thoroughly saturated with the pumping pickle.

Under the present methods of curing, the greater proportion of the sours occur among the partly pumped or mild-cure hams. These hams are pumped in the shank only, and the growth of the ham-souring bacillus within the bodies of these hams is not interfered with until the curing pickle has penetrated from the outside. As it requires several weeks for the curing pickle to penetrate thoroughly into the deeper portions of these hams, the bacillus is thus afforded a considerable interval in which to develop before it is exposed to the inhibitory action of the pickle. If these hams could be thoroughly pumped in the body at the beginning of the cure in the same manner as the regular-cure hams, the chief loss from ham souring would be eliminated. It would not do, however, to pump these hams in the body with the same pumping pickle used in the regular cure, as the meat would be rendered too salty and the mild flavor of the ham would be lost. There is undoubtedly a demand for mild-cure hams, otherwise they would not be on the market; and the question then arises how to pump these hams and still retain a mild cure. This might be accomplished by pumping these hams with their own curing pickle, which is usually a milder pickle than that employed in the regular cure, or an even milder pumping pickle might be used. If mild-cure hams were pumped in this way, the percentage of souring in these hams could undoubtedly be greatly diminished without materially affecting the flavor of the ham.

To recapitulate briefly, the prevention of ham souring is to be sought in two ways:(1) Through greater care in handling the hams and the adoption of precautionary measures to prevent the introduction of the ham-souring bacillus into the bodies of the hams, and (2) through more thorough pumping of the deeper or inner portions of the hams, so as to create an unfavorable soil or medium in which the ham-souring bacillus can not develop even if it should gain entrance into the bodies of the hams.

From what has been said it will be apparent that ham souring can probably never be entirely eliminated from the packing house under the present methods of curing, but the adoption of precautionary measures in testing and pumping hams, together with a more thorough pumping of all hams in ways similar to those suggested, would unquestionably reduce very materially the losses from this source.

GENERAL SUMMARY AND CONCLUSIONS.

1. In this paper it has been shown that ham souring, as encountered in the wet cure where the hams are entirely submerged in pickling fluids, is due to the growth of an anaerobic bacillus within the bodies of the hams. This bacillus (*B. putrefaciens*) was found in sour hams obtained from four different packing establishments. It was isolated and grown in various laboratory media, in one of which, the egg-pork medium, it gave rise to the characteristic sour-ham odor. This bacillus was the only organism that could be isolated from sour hams that was capable of producing the characteristic sour-ham odor in the egg-pork medium.

2. When injected into the bodies of sound hams, the bacillus caused these hams to sour in the process of curing. In hams which had been inoculated with the bacillus and thus artificially soured, the bacillus was recovered in cultures taken at points far removed, relatively speaking, from the point of inoculation, indicating that the bacillus had multiplied and progressed by extension throughout the bodies of the hams.

3. The bacillus possesses no motility, and its extension throughout the bodies of the hams is a result of multiplication. In its growth it follows along the connective-tissue bands between the muscle bundles, which are composed of comparatively loose tissue and afford paths of least resistance. When it invades the muscle tissue proper, it follows along the sarcolemma sheaths between the muscle fibers. As a result of this growth the muscular tissue becomes softer and tends to break more easily.

4. The bacillus belongs to the class of putrefactive anaerobes which are widely distributed in nature in dust, soil, and excrementitious matters. The bacillus or its spores is present in the dust and dirt of packing houses and finds its way into the hams in the various manipulations to which they are subjected.

5. The bacillus or its spores may be introduced into hams on the thermometers used in testing the hams, on the pumping needles, and possibly on the billhooks used in handling the hams. It may also be carried into the hams in the pumping pickle, and may even find its way into the hams from the curing pickle, although infection through the latter channel probably does not often occur.

6. The bacillus develops in the deeper portions of the ham because of the anaerobic conditions there prevailing, and souring is most often encountered, therefore, in the deeper portions of the ham near the bone.

7. A preliminary study of the chemical changes which take place in the process of souring shows that these changes are of a putrefactive nature, and ham souring, as ordinarily encountered, is to be regarded as an incipient putrefaction. Hams which had been artificially soured by injections of culture were compared with sour hams obtained from the packing house, and the putrefactive changes were found to be identical.

8. Hams which have once become sour can never be restored to a sound condition, because of the chemical changes which result from the growth of the bacillus. In other words, the tissues of the ham undergo certain chemical changes in the process of souring, and when these changes have once taken place the tissues can never be restored to a sound condition. The repumping of slightly soured hams with a strong pumping pickle will check further souring, by inhibiting the growth of the bacillus, but will not restore to a sound condition those portions of the ham which have become sour.

9. The salts of the pickling fluids have a marked inhibitory action on the ham-souring bacillus, and sours occur less frequently in regular-cure hams.

10. In regular-cure hams the growth of the ham-souring bacillus is restricted and often completely inhibited as a result of the additional pumping which these hams receive, whereby they are more or less saturated with pickle at the beginning of the cure.

11. If the pumping of regular-cure hams were more thorough and all of the deeper portions of the ham were thoroughly saturated with the pumping pickle, souring could be largely eliminated if not entirely prevented in the hams, as an unfavorable medium or soil would thus be created, in which the ham-souring bacillus could not develop. The reason that souring does develop in regular-cure hams is because the pumping is not always thorough and there are certain areas in the deeper portions of these hams which are not saturated with the pumping pickle.

12. Under the present methods of curing, the partly pumped or mild-cure hams furnish the greater proportion of the sours, as these hams are not pumped in the body and the growth of the ham-souring bacillus within the bodies of these hams is not interfered with until the curing pickle has penetrated from the outside. As it requires several weeks for the curing pickle to penetrate thoroughly into the deeper portions of these hams, the bacillus is thus afforded a considerable interval in which to develop.

13. The percentage of souring in the mild-cure hams could be greatly reduced without materially affecting the cure by pumping these hams with their own curing pickle, which is usually a milder pickle than that employed

in the regular cure; and if the pumping were thorough the number of sours in these hams could be reduced to a small figure.

14. The only way by which ham souring could be entirely eliminated from the larger packing establishments under the present methods of curing would be to handle the hams throughout under aseptic conditions, and this, for obvious reasons, would be an impossibility. The losses from ham souring may be materially reduced, however, by greater care in handling the hams and the adoption of precautionary measures designed to prevent the introduction of contaminated foreign matter into the bodies of the hams, together with more thorough methods of pumping.

ACKNOWLEDGMENTS.

In conclusion, the writer desires to express his obligations to Dr. S. E. Bennett, of the Inspection Division, inspector in charge at Chicago, for the assignment of trained meat inspectors to assist in the work, as well as for kind assistance in obtaining data and material for laboratory study, and to Dr. L. E. Day, of the Pathological Division, who kindly prepared the sections which are figured and described in the present article.